# Mechanics of Functionally Graded Materials and Structures

*Edited by Farzad Ebrahimi*

Published in London, United Kingdom

IntechOpen

*Supporting open minds since 2005*

Mechanics of Functionally Graded Materials and Structures
http://dx.doi.org/10.5772/intechopen.81347
Edited by Farzad Ebrahimi

Contributors
Dongxing Cao, Bin Wang, Wenhua Hu, Yanhui Gao, Mohamed Fahmy, Archana Nigrawal, Nguyen Van Thuan, Cherian Paul, Ramasamy Sellamuthu, Tiejun Liu



First published in London, United Kingdom, 2020 by IntechOpen
IntechOpen is the global imprint of INTECHOPEN LIMITED, registered in England and Wales, registration number: 11086078, 7th floor, 10 Lower Thames Street, London, EC3R 6AF, United Kingdom
Printed in Croatia

British Library Cataloguing-in-Publication Data
A catalogue record for this book is available from the British Library

Additional hard and PDF copies can be obtained from orders@intechopen.com

Mechanics of Functionally Graded Materials and Structures
Edited by Farzad Ebrahimi
p. cm.
Print ISBN 978-1-78984-554-9
Online ISBN 978-1-78984-555-6
eBook (PDF) ISBN 978-1-83968-404-3

For EU product safety concerns:
IN TECH d.o.o., Prolaz Marije Krucifikse Kozulić 3, 51000 Rijeka, Croatia, info@intechopen.com or visit our website at intechopen.com.

# Meet the editor

Dr. Farzad Ebrahimi was born in Qazvin, Iran, in 1979. He graduated in Mechanical Engineering from the University of Tehran, Iran, in 2002. He received his MSc and PhD in Mechanical Engineering, with a specialization in applied design from the University of Tehran, Iran, in 2009. Since 2002, he has been working at the Smart Materials and Structures Lab Research Center of the Faculty of Mechanical Engineering at the University of Tehran, where he is a researcher of smart functionally graded materials and structures. He joined the Department of Mechanical Engineering at the Imam Khomeini International University as an assistant professor in 2010. He is involved in several international journals as editor and reviewer. He serves on the editorial board of the SAGE Publication: "Journal of Advances in Mechanical Engineering". He is the author of the following books: "Smart Functionally Graded Plates" and "Progress in Analysis of Functionally Graded Structures", Nova Science Publishers, NY. He has also served as the editor of the book "Nanocomposites - New Trends and Developments", IntechOpen Access Publisher. His research interests include smart materials and structures, nanostructures, vibration nanocomposites, and composite materials and structures, and he has published several research papers in these fields. His research in these areas has been presented at international conferences and appeared in academic journals such as "Composite Structures", "Composites Part B", "Journal of Mechanical Science and Technology", "Smart Materials and Structures", "European Journal of Mechanics", and "Archive of Applied Mechanics". Dr. Ebrahimi also has a strong collaboration with Iranian industries on gas and oil projects, as well as serving as an ad-hoc referee in several top academic journals.

# Contents

# Preface

Functionally graded materials (FGMs) are inhomogeneous materials, consisting of two (or more) different materials, engineered to have a continuously varying spatial composition profile. This is achieved by gradually varying the volume fraction of the constituent materials. The concept is to produce a composite material by varying the microstructure from one material to another material with a specific gradient. The materials can be designed for specific functions and applications. FGMs offer great promise in applications where the operating conditions are extreme. For example, wear-resistant linings for handling large, heavy, abrasive ore particles; rocket heat shields; heat exchanger tubes; thermoelectric generators; heat-engine components; plasma facings for fusion reactors; and electrically insulating metal/ceramic joints. They are also ideal for minimizing thermomechanical mismatch in metal-ceramic bonding.

This book is a result of contributions of experts from the international scientific community working in different aspects of functionally graded materials and structures and reports on the latest research and development findings on this topic through original and innovative research studies. Through its six chapters, the reader will have access to works related to processing, characteristics, modeling, and applications of functionally graded materials and structures.

The book contains up-to-date publications from leading experts and the edition is intended to provide valuable recent information to the professionals involved in functionally graded materials and structure analysis and applications. The text is addressed not only to researchers, but also to professional engineers, students, and other experts in a variety of disciplines, both academic and industrial seeking to gain a better understanding of what has been done in the field recently, and what open problems are in this area.

I hope that the readers will find this book useful and inspiring by examining the recent developments in functionally graded materials and structures.

Lastly, I would like to thank all the authors for their excellent contributions in different areas covered by this book, and the IntechOpen team, especially the process manager Ms. Ivana Barac, for their support and patience throughout the publishing process.

**Dr. Farzad Ebrahimi**
Department of Mechanical Engineering,
Imam Khomeini International University,
Qazvin, Iran

Chapter 1

# Free Vibration of Axially Functionally Graded Beam

*Dongxing Cao, Bin Wang, Wenhua Hu and Yanhui Gao*

## Abstract

Axially functionally graded (AFG) beam is a special kind of nonhomogeneous functionally gradient material structure, whose material properties vary continuously along the axial direction of the beam by a given distribution form. There are several numerical methods that have been used to analyze the vibration characteristics of AFG beams, but it is difficult to obtain precise solutions for AFG beams because of the variable coefficients of the governing equation. In this topic, the free vibration of AFG beam using analytical method based on the perturbation theory and Meijer G-Function are studied, respectively. First, a detailed review of the existing literatures is summarized. Then, based on the governing equation of the AFG Euler-Bernoulli beam, the detailed analytic equations are derived on basis of the perturbation theory and Meijer G-function, where the nature frequencies are demonstrated. Subsequently, the numerical results are calculated and compared, meanwhile, the analytical results are also confirmed by finite element method and the published references. The results show that the proposed two analytical methods are simple and efficient and can be used to conveniently analyze free vibration of AFG beam.

**Keywords:** axially functionally graded beams, free vibration, natural frequency, asymptotic perturbation method, Meijer G-function, finite segment model

## 1. Introduction

Functionally gradient materials (FGMs) make a composite material by varying the microstructure from one material to another material with a specific gradient. It can be designed for specific function and applications. If it is for thermal or corrosive resistance or malleability and toughness, both strengths of the material may be used to avoid corrosion, fatigue, fracture, and stress corrosion cracking. FGMs are usually made into several structures, such as beams [1–4], plates [5–8], and shells [9–12]. In this area, the variation of material properties in functionally graded beams may be oriented in transverse (thickness) direction or/and longitudinal/axial (length) direction.

For functionally graded beams with thickness-wise gradient variation, there have been many studies devoted to this topic. Lee et al. [13] establish an accurate transfer matrix method to analyze the free vibration characteristics of FGM beams whose Young's modulus and density vary continuously with the height of the beam section through power law distribution. Su et al. [14] developed the dynamic stiffness method to investigate the free vibration behavior of FGM beams. Jing et al. [15]

introduced a new approach by combining the cell-centered finite volume method and Timoshenko beam theory to analyze static and free vibration of FGM beams. Ait Atmane et al. [16] investigated the free vibration of a nonuniform FGM beams with exponentially varying width and material properties. Sina et al. [17] studied the free vibration of FGM beams by analytical method based on the traditional first-order shear deformation theory. Sharma [18] investigated the computational characteristics of harmonic differential quadrature method for free vibration of functionally graded piezoelectric material beam, which the material properties are assumed to have a power law or sigmoid law variation across the depth. Li et al. [19] proposed a high-order shear theory for free vibration of FGM beams with continuously varying material properties under different boundary conditions. Celebi et al. [20] employed the complementary function method to investigate the free vibration analysis of simply supported FGM beams, which the material properties change arbitrarily in the thickness direction. Chen et al. [21] studied the nonlinear free vibration behavior of shear deformable sandwich porous FGM beam based on the von kármán type geometric nonlinearity and Ritz method. Nazemnezhad and Hosseini-Hashemi [22] examined the nonlinear free vibration of FGM nanobeams with immovable ends using the multiple scale method.

As the FGMs are good for severe conditions, thermal-mechanical effect on FGM structures has attracted broad attention. In this field, Farzad Ebrahimi and Erfan Salari obtained outstanding achievements. Considering the thermal-mechanical effect and size-dependent thermo-electric effect, the buckling and vibration behavior of FGM nanobeams are studied [23–26]. Considering the concept of neutral axis, they [27] studied the free buckling and vibration of FGM nanobeams using semi-analytical differential transformation method. To discuss the effect of the shear stress, Reddy's higher-order shear deformation beam theory is introduced to study the vibration of the FGM structures [28–30]. Ebrahimi et al. [31–33] also studied vibration characteristics of FGM beams with porosities. Based on nonlocal elasticity theory, the nonlocal temperature-dependent vibration of FGM nanobeams were studied in thermal environment [34–36].

Another significant class of functionally graded beams is those with lengthwise varying material properties. It is difficult to obtain precise solutions for axially functionally graded (AFG) beams because of the variable coefficients of the governing equation. To solve this problem, a great deal of methods has been used to analyze the vibration characteristics of AFG beams. By assuming that the material constituents vary throughout the longitudinal directions according to a simple power law, Alshorbagy et al. [37] developed a two-node, six-degree-of-freedom finite element method (FEM) in conjunction with Euler-Bernoulli beam theory to detect the free vibration characteristics of a functionally graded beam. Shahba et al. [38, 39] used the FEM to study the free vibration of an AFG-tapered beam based on Euler-Bernoulli and Timoshenko beam theory. Shahba and Rajasekaran [40] studied the free vibration analysis of AFG-tapered Euler-Bernoulli beams employing the differential transform element method. Liu et al. [41] applied the spline finite point method to investigate the same problems. Rajasekaran [42] researched the free bending vibration of rotating AFG-tapered Euler-Bernoulli beams with different boundary conditions using the differential transformation method and differential quadrature element method. Rajasekaran and Tochaei [43] carried out the free vibration analysis of AFG Timoshenko beams using the same method. Huang and Li [44] studied the free vibration of variable cross-sectional AFG beams. The differential equation with variable coefficients is combined with the boundary conditions and transformed into Fredholm integral equation. By solving Fredholm integral equation, the natural frequencies of AFG beams can be obtained. Huang et al. [45] proposed a new approach for investigating the vibration behaviors of AFG

Timoshenko beams with nonuniform cross section by introducing an auxiliary function. Huang and Rong [46] introduced a simple approach to deal with the free vibration of nonuniform AFG Euler-Bernoulli beams based on the polynomial expansion and integral technique. Hein and Feklistova [47] solved the vibration problems of AFG beams with various boundary conditions and varying cross sections via the Haar wavelet series. Xie et al. [48] presented a spectral collocation approach based on integrated polynomials combined with the domain decomposition technique for free vibration analyses of beams with axially variable cross sections, moduli of elasticity, and mass densities. Kukla and Rychlewska [49] proposed a new approach to study the free vibration analysis of an AFG beam; the approach relies on replacing functions characterizing functionally graded beams with piecewise exponential functions. Zhao et al. [50] introduced a new approach based on Chebyshev polynomial theory to investigate the free vibration of AFG Euler-Bernoulli and Timoshenko beams with nonuniform cross sections. Fang and Zhou [51, 52] researched the modal analysis of rotating AFG-tapered Euler-Bernoulli and Timoshenko beams with various boundary conditions employing the Chebyshev-Ritz method. Li et al. [53, 54] obtained the exact solutions for the free vibration of FGM beams with material profiles and cross-sectional parameters varying exponentially in the axial direction, where assumptions of Euler-Bernoulli and Timoshenko beam theories have been applied, respectively. Sarkar and Ganguli [55] studied the free vibration of AFG Timoshenko beams with different boundary conditions and uniform cross sections. Akgöz and Civalek [56] examined the free vibrations of AFG-tapered Euler-Bernoulli microbeams based on Bernoulli-Euler beam and modified couple stress theory. Yuan et al. [57] proposed a novel method to simplify the governing equations for the free vibration of Timoshenko beams with both geometrical nonuniformity and material inhomogeneity along the beam axis, and a series of exact analytical solutions are derived from the reduced equations for the first time. Yilmaz and Evran [58] investigated the free vibration of axially layered FGM short beams using experimental and FEM simulation, which the beams are manufactured by using the powder metallurgy technique using different weight fractions of aluminum and silicon carbide powders.

Till now, there also are plenty of literatures devoted to the free vibration for nonuniform beams. Boiangiu et al. [59] obtained the exact solutions for free bending vibrations of straight beams with variable cross section using Bessel's functions and proposed a transfer matrix method to determine the natural frequencies of a complex structure of conical and cylindrical beams. Garijo [60] analyzed the free vibration of Euler-Bernoulli beams of variable cross section employing a collocation technique based on Bernstein polynomials. Arndt et al. [61] presented an adaptive generalized FEM to determine the natural frequencies of nonuniform Euler-Bernoulli beams. The spline-method of degree 5 defect 1 is proposed by Zhernakov et al. [62] to determine the natural frequencies of beam with variable cross section. Wang [63] studied the vibration of a cantilever beam with constant thickness and linearly tapered sides by means of a novel accurate, efficient initial value numerical method. Silva and Daqaq [64] solved the linear eigenvalue problem exactly of a slender cantilever beam of constant thickness and linearly varying width using the Meijer G-function approach. Rajasekaran and Khaniki [65] applied the FEM to research the vibration behavior of nonuniform small-scale beams in the framework of nonlocal strain gradient theory. Çalım [66] investigated the dynamic behavior of nonuniform composite beams employing an efficient method of analysis in the Laplace domain. Yang et al. [67] applied the power series method to investigate the natural frequencies and the corresponding complex mode functions of a rotating tapered cantilever Timoshenko beam. Clementi [68] analytically determined the frequency response curves of a nonuniform beam undergoing nonlinear oscillations by the multiple time scale method. Wang [69] proposed the differential quadrature element method for the natural frequencies of multiple-stepped beams with an aligned neutral axis. Abdelghany [70] utilized the differential transformation method to examine the free vibration of nonuniform circular beam.

The asymptotic development method, which is a kind of perturbation analysis method, is always used to solve nonlinear vibration equations. For example, Chen et al. [71, 72] studied the nonlinear dynamic behavior of axially accelerated visco-elastic beams and strings based on the asymptotic perturbation method. Ding et al. [73, 74] studied the influence of natural frequency of transverse vibration of axially moving viscoelastic beams and the steady-state periodic response of forced vibration of dynamic viscoelastic beams based on the multi-scale method. Chen [75] used the asymptotic perturbation method to analyze the finite deformation of prestressing hyperelastic compression plates. Hao et al. [76] employed the asymptotic perturbation method to obtain the nonlinear dynamic responses of a cantilever FGM rectangular plate subjected to the transversal excitation in thermal environment. Andrianov and Danishevs'kyy [77] proposed an asymptotic method for solving periodic solutions of nonlinear vibration problems of continuous structures. Based on the asymptotic expansion method of Poincaré-Lindstedt version [78], the longitudinal vibration of a bar and the transverse vibration of a beam under the action of a nonlinear restoring force are studied. The asymptotic development method is applied to obtain an approximate analytical expression of the natural frequencies of nonuniform cables and beams [79, 80]. Cao et al. [81, 82] applied the asymptotic development method to analyze the free vibration of nonuniform axially functionally graded (AFG) beams. Tarnopolskaya et al. [83] gave the first comprehensive study of the mode transition phenomenon in vibration of beams with arbitrarily varying curvature and cross section on the basis of asymptotic analysis.

The present topic focus on the free vibration of AFG beams with uniform or nonuniform cross sections using analytical method: the asymptotic perturbation method (APM) and Meijer G-function. First, the governing differential equation for free vibration of nonuniform AFG beam is summarized and rewritten in a form of a dimensionless equation based on Euler-Bernoulli beam theory. Second, the analytic equations are then derived in detail in Sections 3 and 4, respectively, where the nature frequencies are obtained and compared with the results of the finite element method and the published references. Finally, the conclusions are presented.

## 2. Governing equation of the AFG beam

This studied free vibration of the axially functionally graded beam, which is a nonuniform and nonhomogeneous structure because of the variable width and height, as shown in **Figure 1**. Based on Euler-Bernoulli beam theory, the governing differential equation of the beam can be written as

$$\frac{\partial^2}{\partial x^2}\left[E(x)I(x)\frac{\partial^2 w(x,t)}{\partial x^2}\right]+\rho(x)A(x)\frac{\partial^2 w(x,t)}{\partial t^2}=0,\qquad 0\leq x\leq L \tag{1}$$

where $w(x,t)$ is the transverse deflection at position $x$ and time $t$; $L$ is the length of the beams; $E(x)I(x)$ is the bending stiffness, which is determined by Young's modulus $E(x)$ and the area moment of inertia $I(x)$; and $\rho(x)A(x)$ is the unit mass

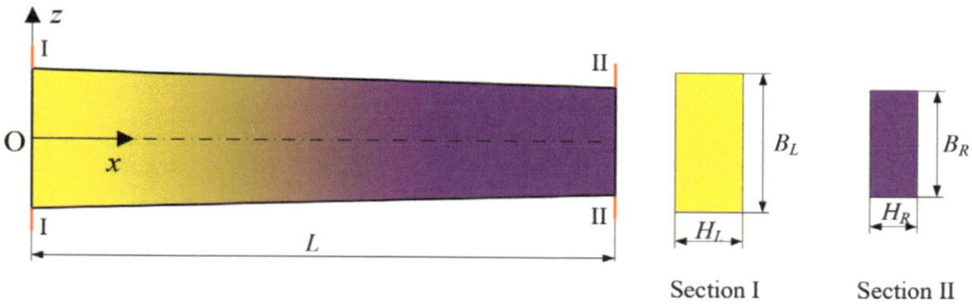


**Figure 1.**
*The geometry and coordinate system of an AFG beam.*

length of beam, which is determined by volume mass density $\rho(x)$ and variable cross-sectional area $A(x)$.

Because of the particularity of AFG beam, bending stiffness $E(x)I(x)$ and unit mass $\rho(x)A(x)$ will change with the axis coordinates, which makes the original constant coefficient differential equation become variable coefficient differential equation and to some extent increases the difficulty of solving. In order to facilitate calculation, we simplify the calculation process by introducing dimensionless parameters. Reference flexural stiffness $EI_0$ and reference mass $\rho A_0$ are introduced, and the above two dimensionless parameters are invariant. Suppose $E(x)I(x) = EI_0 + \overline{E(x)I(x)}$ and $\rho(x)A(x) = \rho A_0 + \overline{\rho(x)A(x)}$, where $EI_0$ and $\rho A_0$ are the invariant parts and $\overline{E(x)I(x)}$ and $\overline{\rho(x)A(x)}$ represent flexural stiffness and mass per unit length, respectively, and vary with the axial coordinates. Here, we introduce a dimensionless space variable $\xi = x/L$ and a dimensionless time variable $\tau = \frac{t}{L^2}\sqrt{\frac{EI_0}{\rho A_0}}$; Eq. (1) can be rewritten in the dimensionless form:

$$\frac{\partial^2}{\partial \xi^2}\left\{[1+f_1(\xi)]\frac{\partial^2 w(\xi,\tau)}{\partial \xi^2}\right\} + [1+f_2(\xi)]\frac{\partial^2 w(\xi,\tau)}{\partial \tau^2} = 0, \qquad 0 \le \xi \le 1 \tag{2}$$

where

$$f_1(\xi) = \frac{\overline{E(\xi)I(\xi)}}{EI_0} \text{ and } f_2(\xi) = \frac{\overline{\rho(\xi)A(\xi)}}{\rho A_0} \tag{3}$$

are the nondimensional varying parts of the flexural stiffness and of the mass per unit length, respectively.

## 3. Asymptotic perturbation method

### 3.1 Equation deriving

In this section, the APM is introduced to obtain a simple proximate formula for the nature frequency of the AFG beam. Firstly, we assume that

$$w(\xi,\tau) = W(\xi)\sin(\omega\tau) \tag{4}$$

where $W(\xi)$ is the amplitude of vibration and $\omega$ is the circular frequency of vibration. We obtain the following equation by substituting Eq. (4) with Eq. (2):

$$\frac{d^2}{d\xi^2}\left\{[1+f_1(\xi)]\frac{d^2W}{d\xi^2}\right\}-\omega^2[1+f_2(\xi)]W=0,\qquad 0\le\xi\le1 \tag{5}$$

To use the APM, a small perturbation parameter $\varepsilon$ is introduced:

$$f_1(\xi)\to\varepsilon f_1(\xi),\ f_2(\xi)\to\varepsilon f_2(\xi) \tag{6}$$

According to the Poincaré-Lindstedt method [78–82], we assume the expansion for $\omega$ and $W(\xi)$ as

$$\begin{aligned}\omega&=\omega_0+\varepsilon\omega_1+\varepsilon^2\omega_2+\ldots,\\ W(\xi)&=W_0(\xi)+\varepsilon W_1(\xi)+\varepsilon^2W_2(\xi)+\ldots.\end{aligned} \tag{7}$$

Substituting these expressions with governing Eq. (5) and then expanding the expressions into a $\varepsilon$-series, Eqs. (8) and (9) are obtained by equating the coefficients of $\varepsilon^0$ and $\varepsilon^1$ to zero, yielding a sequence of problems for the unknowns $\omega_i$ and $W_i(\xi)$:

$$\frac{d^4W_0}{d\xi^4}-\omega_0^2W_0=0 \tag{8}$$

$$\frac{d^4W_1}{d\xi^4}-\omega_0^2W_1+h_1(\xi)-2\omega_1\omega_0W_0=0 \tag{9}$$

where

$$h_1(\xi)=2\frac{df_1(\xi)}{d\xi}\frac{d^3W_0}{d\xi^3}+\frac{d^2f_1(\xi)}{d\xi^2}\frac{d^2W_0}{d\xi^2}+\omega_0^2[f_1(\xi)-f_2(\xi)]W_0 \tag{10}$$

For Eq. (8), the following general solution can be obtained:

$$W_0=A\sin(k\xi)+B\cos(k\xi)+C\sinh(k\xi)+D\cosh(k\xi) \tag{11}$$

where

$$k=\sqrt{\omega_0} \tag{12}$$

For simplicity, we consider clamped-free (C-F) beams, and the corresponding boundary conditions are

$$W_0=\frac{dW_0}{d\xi}=0,\quad \xi=0 \tag{13}$$

$$\frac{d^2W_0}{d\xi^2}=\frac{d^3W_0}{d\xi^3}=0,\quad \xi=1 \tag{14}$$

Then, the following equations from equation can be obtained:

$$\begin{aligned}&A+C=0\\ &B+D=0\\ &\frac{C}{D}=\frac{\sin k-\sinh k}{\cos k+\cosh k}\end{aligned} \tag{15}$$

and the frequency equation is

$$\cos k \cosh k + 1 = 0 \tag{16}$$

The spatial mode shape can be obtained as

$$W_0 = \cosh(k\xi) - \cos(k\xi) + \frac{C}{D}[\sinh(k\xi) - \sin(k\xi)] \tag{17}$$

Now, the solution of the first-order equation is analyzed. In Eq. (9), both $h_1(\xi)$ and $W_1$ are linearly correlated with $W_0$. Based on the theory of ordinary differential equations [84], the solvability conditions of linear differential equations can be expressed by the orthogonality of solutions of homogeneous systems of equations. At the same time, according to the orthogonality of modal vibration theory, the solution of Eq. (9) exists under the condition of the solvability of differential equation:

$$\int_0^1 [h_1(\xi) - 2\omega_1\omega_0 W_0] W_0 d\xi = 0 \tag{18}$$

is satisfied. As a result,

$$\omega_1 = \frac{\int_0^1 h_1(\xi) W_0 d\xi}{2\omega_0 \int_0^1 W_0^2 d\xi} \tag{19}$$

Because $h_1(\xi)$ is linearly correlated with $W_0$, the former equations indicate that the arbitrary amplitude of $W_0$ does not impact $\omega_1$. This finding yields the first-order correction of the natural frequency $\omega_0$ corresponding to a nonuniform and homogeneous beam.

Integrating by parts, we obtain

$$\begin{aligned}\int_0^1 h_1(\xi) W_0 d\xi = &\left(\frac{df_1}{d\xi}\frac{d^2W_0}{d\xi^2}W_0 + f_1\frac{d^3W_0}{d\xi^3}W_0 - f_1\frac{d^2W_0}{d\xi^2}\frac{dW_0}{d\xi}\right)\Bigg|_0^1 \\ &+ \int_0^1 \left[f_1\left(\frac{d^2W_0}{d\xi^2}\right)^2 - \omega_0^2 f_2 W_0^2\right] d\xi\end{aligned} \tag{20}$$

By definition we have

$$f_1(\xi) = \frac{\overline{E(\xi)I(\xi)}}{E_0 I} = \frac{E(\xi)I(\xi) - EI_0}{EI_0} \tag{21}$$

so that

$$\begin{aligned}&\left(\frac{df_1}{d\xi}\frac{d^2W_0}{d\xi^2}W_0 + f_1\frac{d^3W_0}{d\xi^3}W_0 - f_1\frac{d^2W_0}{d\xi^2}\frac{dW_0}{d\xi}\right)\Bigg|_0^1 + \int_0^1 f_1\left(\frac{d^2W_0}{d\xi^2}\right)^2 d\xi \\ &= \left\{\frac{d[E(\xi)I(\xi)]}{EI_0 d\xi}\frac{d^2W_0}{d\xi^2}W_0 + \frac{E(\xi)I(\xi)}{EI_0}\frac{d^3W_0}{d\xi^3}W_0 - \frac{E(\xi)I(\xi)}{EI_0}\frac{d^2W_0}{d\xi^2}\frac{dW_0}{d\xi}\right. \\ &\left. + \frac{d^2W_0}{d\xi^2}\frac{dW_0}{d\xi} - \frac{d^3W_0}{d\xi^3}W_0\right\}\Bigg|_0^1 + \frac{1}{EI_0}\int_0^1 E(\xi)I(\xi)\left(\frac{d^2W_0}{d\xi^2}\right)^2 d\xi - \int_0^1 \left(\frac{d^2W_0}{d\xi^2}\right)^2 d\xi\end{aligned} \tag{22}$$

Choosing the reference bending stiffness

$$EI_0 = \frac{\left\{ \frac{d[E(\xi)I(\xi)]}{d\xi} \frac{d^2 W_0}{d\xi^2} W_0 + E(\xi)I(\xi) \frac{d^3 W_0}{d\xi^3} W_0 - E(\xi)I(\xi) \frac{d^2 W_0}{d\xi^2} \frac{dW_0}{d\xi} \right\} \Big|_0^1 + \int_0^1 E(\xi)I(\xi) \left( \frac{d^2 W_0}{d\xi^2} \right)^2 d\xi}{\left( \frac{d^3 W_0}{d\xi^3} W_0 - \frac{d^2 W_0}{d\xi^2} \frac{dW_0}{d\xi} \right) \Big|_0^1 + \int_0^1 \left( \frac{d^2 W_0}{d\xi^2} \right)^2 d\xi} \tag{23}$$

we have $\left( \frac{df_1}{d\xi} \frac{d^2 W_0}{d\xi^2} W_0 + f_1 \frac{d^3 W_0}{d\xi^3} W_0 - f_1 \frac{d^2 W_0}{d\xi^2} \frac{dW_0}{d\xi} \right) \Big|_0^1 + \int_0^1 f_1 \left( \frac{d^2 W_0}{d\xi^2} \right)^2 d\xi = 0.$

Analogously, we choose

$$\rho A_0 = \frac{\int_0^1 \rho(\xi) A(\xi) W_0^2 d\xi}{\int_0^1 W_0^2 d\xi} \tag{24}$$

giving $\int_0^1 f_2 W_0^2 d\xi = 0$. Then, we obtain $\omega_1 = 0$. These values are the properties of the equivalent homogeneous beam having the same frequency (at least up to the first order) as the given nonuniform AFG beam.

The nth natural circular frequency of the AFG beam can be derived as

$$\lambda_n = \frac{1}{L^2} \sqrt{\frac{EI_0}{\rho A_0}} \omega_0 \tag{25}$$

Each order of frequency of $\omega_0$ can be determined by Eq. (16) (in turn, positive numbers from small to large). The required variables have been computed by the above expression. Eq. (25) is an approximate formula for the natural frequencies of variable cross-sectional AFG beams.

In order to show the applicability of this method, we study other supporting conditions, and we can easily get the corresponding boundary conditions of Eqs. (13) and (14). Due to the limited space, the detailed derivation process is omitted, and the final results are shown in **Table 1**.

### 3.2 Numerical results and discussion

Based on the above analysis, four kinds of AFG beams with various taper ratios are considered, as shown in **Figure 2**. The numerical simulations are carried out, and the results are compared with the published literature results to verify the validity of the proposed method.

In **Figure 2**, $B_L$ and $B_R$ are the width of the left and right ends of the beams, respectively, and $H_L$ and $H_R$ are the height of the left and right ends of the beams, respectively. Here, we assume that the geometric characteristics of AFG beams vary linearly along the longitudinal direction. Therefore, the variation of

| Boundary conditions | Frequency equation | Mode shape |
|---|---|---|
| Simply supported (S-S) | $\sin k = 0$ | $W_0 = \sin(k\xi)$ |
| Clamped-pinned (C-P) | $\tan k - \tanh k = 0$ | $W_0 = \cosh(k\xi) - \cos(k\xi) - \frac{\cosh k - \cos k}{\sinh k - \sin k}[\sinh(k\xi) - \sin(k\xi)]$ |
| Clamped-clamped (C-C) | $\cos k \cosh k - 1 = 0$ | $W_0 = \cosh(k\xi) - \cos(k\xi) + \frac{\sin k + \sinh k}{\cos k - \cosh k}[\sinh(k\xi) - \sin(k\xi)]$ |

**Table 1.**
*Frequency equations and mode shapes for various beams.*

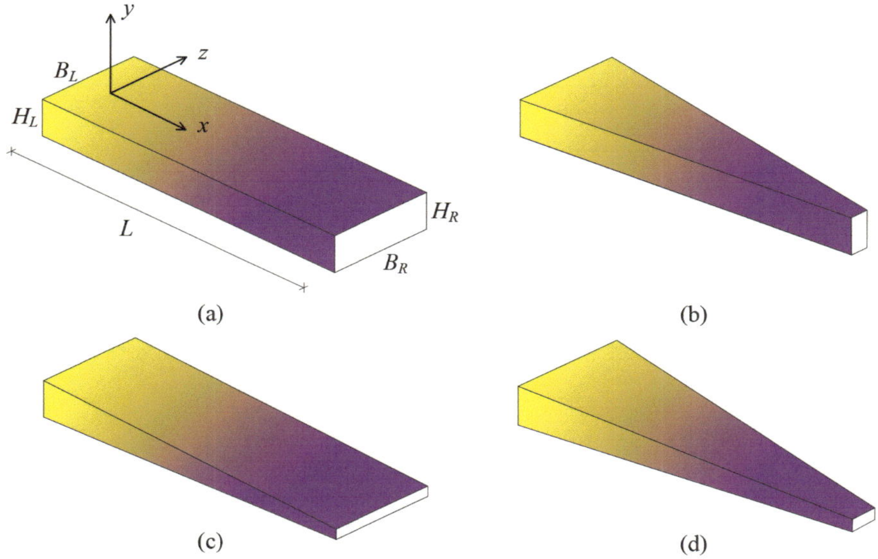


**Figure 2.**
*Geometry and coordinate system of an AFG beam for different taper ratios: (a) case 1, $c_b = c_h = 0$; (b) case 2, $c_h = 0, c_b \neq 0$; (c) case 3, $c_b = 0, c_h \neq 0$; and (d) case 4, $c_b \neq 0, c_h \neq 0$.*

cross-sectional area $A(x)$ and moment of inertia $I(x)$ along the beam axis can be clearly expressed as follows:

$$A(x) = A_L\left(1 - c_b\frac{x}{L}\right)\left(1 - c_h\frac{x}{L}\right),\ \ I(x) = I_L\left(1 - c_b\frac{x}{L}\right)\left(1 - c_h\frac{x}{L}\right)^3 \tag{26}$$

where $c_b = 1 - \frac{B_R}{B_L}$ and $c_h = 1 - \frac{H_R}{H_L}$ are the breadth and height taper ratios, respectively. $A_L$ and $I_L$ are cross-sectional area and area moment of inertia of the beam left sides, respectively. It is instructive to remember that if $c_b = c_h = 0$, the beam would be uniform; if $c_h = 0, c_b \neq 0$, the beam would be tapered with constant height; if $c_b = 0, c_h \neq 0$, the beam would be tapered with constant width; and if $c_b \neq 0, c_h \neq 0$, the beam would be double tapered. These four cases are corresponding to **Figure 2(a)–(d)**, respectively. Moreover, the material properties such as Young's modulus $E(x)$ and mass density $\rho(x)$ along the beam axis are assumed as

$$E(x) = E_L\left(1 + \frac{x}{L}\right),\ \ \rho(x) = \rho_L\left[1 + \frac{x}{L} + \left(\frac{x}{L}\right)^2\right] \tag{27}$$

where $E_L$ and $\rho_L$ are Young's modulus and mass density of the beam left sides, respectively.

Based on the introduced analytical equation, the first third-order nondimensional natural frequencies ($\Omega_n = \lambda_n L^2\sqrt{\rho_L A_L / E_L I_L}$) of the four cases of nonuniform AFG beams with different boundary configurations were obtained. The results were listed in **Tables 2–7**, respectively, and it also was compared with those of published work by Shahba et al. [38].

**Table 2** shows the first third-order natural frequencies of the AFG beam, case of **Figure 2(a)**, which is uniform but nonhomogeneous. It can be clearly seen that the analytical results obtained from the asymptotic development method are in good agreement with those given by Ref. [38].

| Boundary condition | | First mode | Second mode | Third mode |
|---|---|---|---|---|
| C-F | Present | 2.439 | 18.437 | 54.339 |
| | Ref. [38] | 2.426 | 18.604 | 55.180 |
| S-S | Present | 9.053 | 35.834 | 80.470 |
| | Ref. [38] | 9.029 | 36.372 | 81.732 |
| C-C | Present | 20.585 | 56.251 | 109.869 |
| | Ref. [38] | 20.472 | 56.549 | 110.947 |

**Table 2.**
*Nondimensional natural frequencies of the AFG uniform beam (case 1) with different boundary conditions.*

| $c_b$ | | C-F | | | S-S | | | C-C | | |
|---|---|---|---|---|---|---|---|---|---|---|
| | | First mode | Second mode | Third mode | First mode | Second mode | Third mode | First mode | Second mode | Third mode |
| 0.2 | Present | 2.613 | 18.887 | 54.951 | 9.068 | 35.957 | 80.772 | 20.457 | 56.196 | 110.003 |
| | Ref. [38] | 2.605 | 19.004 | 55.534 | 9.060 | 36.342 | 81.685 | 20.415 | 56.472 | 110.862 |
| 0.4 | Present | 2.854 | 19.483 | 55.753 | 9.088 | 36.117 | 81.165 | 20.294 | 56.124 | 110.177 |
| | Ref. [38] | 2.851 | 19.530 | 56.023 | 9.087 | 36.315 | 81.645 | 20.288 | 56.298 | 110.671 |
| 0.6 | Present | 3.214 | 20.311 | 56.853 | 9.113 | 36.332 | 81.697 | 20.079 | 56.028 | 110.411 |
| | Ref. [38] | 3.214 | 20.296 | 56.800 | 9.099 | 36.297 | 81.624 | 20.019 | 55.921 | 110.250 |
| 0.8 | Present | 3.832 | 21.542 | 58.453 | 9.147 | 36.638 | 82.456 | 19.783 | 55.892 | 110.743 |
| | Ref. [38] | 3.831 | 21.676 | 58.435 | 9.069 | 36.277 | 81.639 | 19.385 | 54.971 | 109.142 |

**Table 3.**
*Nondimensional natural frequencies of the AFG-tapered beam with constant height (case 2) and different boundary conditions.*

| $c_h$ | | C-F | | | S-S | | | C-C | | |
|---|---|---|---|---|---|---|---|---|---|---|
| | | First mode | Second mode | Third mode | First mode | Second mode | Third mode | First mode | Second mode | Third mode |
| 0.2 | Present | 2.5054 | 17.2596 | 49.4982 | 8.1416 | 32.1888 | 72.2680 | 18.2420 | 50.1851 | 98.2992 |
| | Ref. [38] | 2.5051 | 17.3802 | 50.0491 | 8.1341 | 32.5236 | 73.1138 | 18.2170 | 50.4801 | 99.1734 |
| 0.4 | Present | 2.6293 | 16.2995 | 45.3519 | 7.2793 | 28.9717 | 65.1203 | 16.3027 | 45.0600 | 88.4345 |
| | Ref. [38] | 2.6155 | 16.0705 | 44.6181 | 7.1531 | 28.4747 | 63.9942 | 15.8282 | 44.0246 | 86.6272 |
| 0.6 | Present | 2.8535 | 15.6697 | 42.2358 | 6.4872 | 26.3694 | 59.4850 | 14.9152 | 41.2502 | 80.9747 |
| | Ref. [38] | 2.7835 | 14.6508 | 38.7446 | 6.0357 | 24.1101 | 54.0921 | 13.2293 | 36.9653 | 72.8740 |
| 0.8 | Present | 3.2889 | 15.5662 | 40.6554 | 5.7966 | 24.6371 | 55.9734 | 14.2233 | 39.1823 | 76.7690 |
| | Ref. [38] | 3.0871 | 13.1142 | 32.1309 | 4.6520 | 19.1314 | 42.6954 | 10.2235 | 28.7492 | 56.8109 |

**Table 4.**
*Nondimensional natural frequencies of the AFG-tapered beam with constant width (case 3) and different boundary conditions.*

| $c_b$ | | 0.2 | 0.4 | 0.6 | 0.8 |
|---|---|---|---|---|---|
| $c_h$ | First mode | | | | |
| 0.2 | Present | 2.6873 | 2.9380 | 3.3113 | 3.9455 |
| | Ref. [38] | 2.6863 | 2.9336 | 3.2993 | 3.9219 |
| 0.4 | Present | 2.8226 | 3.0877 | 3.4796 | 4.1377 |
| | Ref. [38] | 2.7987 | 3.0486 | 3.4181 | 4.0471 |
| 0.6 | Present | 3.0640 | 3.3506 | 3.7700 | 4.4625 |
| | Ref. [38] | 2.9699 | 3.2237 | 3.5985 | 4.2355 |
| 0.8 | Present | 3.5271 | 3.8475 | 4.3081 | 5.0458 |
| | Ref. [38] | 3.2794 | 3.5401 | 3.9232 | 4.5695 |
| $c_h$ | Second mode | | | | |
| 0.2 | Present | 17.7225 | 18.3289 | 19.1598 | 20.3725 |
| | Ref. [38] | 17.7501 | 18.2379 | 18.9501 | 20.2432 |
| 0.4 | Present | 16.7822 | 17.4061 | 18.2458 | 19.4418 |
| | Ref. [38] | 16.4092 | 16.8571 | 17.5139 | 18.7164 |
| 0.6 | Present | 16.1771 | 16.8214 | 17.6687 | 18.8380 |
| | Ref. [38] | 14.9567 | 15.3627 | 15.9616 | 17.0694 |
| 0.8 | Present | 16.0947 | 16.7493 | 17.5836 | 18.6877 |
| | Ref. [38] | 13.3850 | 13.7466 | 14.2848 | 15.2955 |
| $c_h$ | Third mode | | | | |
| 0.2 | Present | 50.2194 | 51.1534 | 52.4114 | 54.1995 |
| | Ref. [38] | 50.3934 | 50.8645 | 51.6029 | 53.1332 |
| 0.4 | Present | 46.1970 | 47.2734 | 48.6925 | 50.6520 |
| | Ref. [38] | 44.9504 | 45.4003 | 46.0957 | 47.5129 |
| 0.6 | Present | 43.2042 | 44.4117 | 45.9613 | 48.0269 |
| | Ref. [38] | 39.0605 | 39.4844 | 40.1304 | 41.4236 |
| 0.8 | Present | 41.7065 | 42.9817 | 44.5636 | 46.5828 |
| | Ref. [38] | 32.4229 | 32.8123 | 33.3986 | 34.5521 |

**Table 5.**
*Nondimensional natural frequencies of the AFG double-tapered beam (case 4); boundary conditions: C-F.*

As can be seen from **Tables 3** and **4**, the first third-order dimensionless natural frequencies of AFG conical beams with only varying width or height are studied, respectively. It is easy to find the following conclusions. This method has higher accuracy on the equal height AFG-tapered beam. When the height changes, there is a certain fractional error in the AFG-tapered beam.

According to **Figure 2(d)**, when the height and width of AFG beams change simultaneously, we can see that AFG beams are not uniform. The natural frequencies of three boundary conditions (free clamping, simply supported, and clamping) are studied in **Tables 5–7**. From the data in the table, it can be clearly found that the natural frequencies of AFG beams at low order are in good agreement with Ref. [38], while at high order, there are some errors in the natural frequencies.

| $c_b$ | | 0.2 | 0.4 | 0.6 | 0.8 |
|---|---|---|---|---|---|
| $c_h$ | First mode | | | | |
| 0.2 | Present | 8.1682 | 8.2018 | 8.2456 | 8.3051 |
| | Ref. [38] | 8.1462 | 8.1498 | 8.1336 | 8.0646 |
| 0.4 | Present | 7.3172 | 7.3647 | 7.4262 | 7.5089 |
| | Ref. [38] | 7.1455 | 7.1254 | 7.0794 | 6.9703 |
| 0.6 | Present | 6.5357 | 6.5960 | 6.6732 | 6.7754 |
| | Ref. [38] | 6.0082 | 5.9638 | 5.8868 | 5.7351 |
| 0.8 | Present | 5.8537 | 5.9240 | 6.0128 | 6.1283 |
| | Ref. [38] | 4.6046 | 4.5355 | 4.4264 | 4.2283 |
| $c_h$ | Second mode | | | | |
| 0.2 | Present | 32.4133 | 32.7007 | 33.0819 | 33.6118 |
| | Ref. [38] | 32.5123 | 32.5079 | 32.5164 | 32.5326 |
| 0.4 | Present | 29.2971 | 29.7076 | 30.2419 | 30.9665 |
| | Ref. [38] | 28.4822 | 28.5003 | 28.5370 | 28.5928 |
| 0.6 | Present | 26.7834 | 27.2965 | 27.9493 | 28.8091 |
| | Ref. [38] | 24.1371 | 24.1791 | 24.2469 | 24.3497 |
| 0.8 | Present | 25.1032 | 25.6683 | 26.3683 | 27.2590 |
| | Ref. [38] | 19.1803 | 19.2509 | 19.3590 | 19.5300 |
| $c_h$ | Third mode | | | | |
| 0.2 | Present | 72.8179 | 73.5237 | 74.4625 | 75.7732 |
| | Ref. [38] | 73.0959 | 73.0903 | 73.1116 | 73.1855 |
| 0.4 | Present | 65.9158 | 66.9202 | 68.2291 | 70.0069 |
| | Ref. [38] | 64.0054 | 64.0350 | 64.1007 | 64.2374 |
| 0.6 | Present | 60.4922 | 61.7392 | 63.3243 | 65.4089 |
| | Ref. [38] | 54.1330 | 54.1992 | 54.3126 | 54.5207 |
| 0.8 | Present | 57.0969 | 58.4547 | 60.1303 | 62.2530 |
| | Ref. [38] | 42.7677 | 42.8742 | 43.0436 | 43.3451 |

**Table 6.**
*Nondimensional natural frequencies of the AFG double-tapered beam (case 4); boundary conditions: S-S.*

## 4. The method of Meijer G-function

### 4.1 Equation deriving

In this section, the Meijer G-function is introduced to obtain the formula of the nature frequency of the AFG beam. Here, a special case of AFG beam is considered, where the cross section is uniform. Thus, in Eq. (1), Young's modulus $E(x)$ and material mass density $\rho(x)$ are variable parameters, and the area moment of inertia $I$ and the cross-sectional area $A$ are invariant. To solve the governing equation, two parameters are firstly introduced to depict the functional gradient parameter equation:

| $c_b$ | | 0.2 | 0.4 | 0.6 | 0.8 |
|---|---|---|---|---|---|
| $c_h$ | First mode | | | | |
| 0.2 | Present | 18.2779 | 18.3231 | 18.3818 | 18.4612 |
| | Ref. [38] | 18.1996 | 18.1286 | 17.9437 | 17.4566 |
| 0.4 | Present | 16.4975 | 16.7396 | 17.0484 | 17.4563 |
| | Ref. [38] | 15.8498 | 15.8350 | 15.7367 | 15.4025 |
| 0.6 | Present | 15.2512 | 15.6622 | 16.1771 | 16.8423 |
| | Ref. [38] | 13.2896 | 13.3319 | 13.3238 | 13.1529 |
| 0.8 | Present | 14.6662 | 15.2004 | 15.8587 | 16.6925 |
| | Ref. [38] | 10.3229 | 10.4255 | 10.5168 | 10.5339 |
| $c_h$ | Second mode | | | | |
| 0.2 | Present | 50.4430 | 50.7713 | 51.2035 | 51.7981 |
| | Ref. [38] | 50.4565 | 50.3599 | 50.1017 | 49.3728 |
| 0.4 | Present | 45.6257 | 46.3346 | 47.2495 | 48.4763 |
| | Ref. [38] | 44.0553 | 44.0370 | 43.9027 | 43.4066 |
| 0.6 | Present | 42.0890 | 43.1214 | 44.4245 | 46.1234 |
| | Ref. [38] | 37.0509 | 37.1137 | 37.1104 | 36.8678 |
| 0.8 | Present | 40.2151 | 41.4614 | 42.9975 | 44.9420 |
| | Ref. [38] | 28.8912 | 29.0409 | 29.1842 | 29.2402 |
| $c_h$ | Third mode | | | | |
| 0.2 | Present | 98.2992 | 99.7466 | 100.8219 | 102.313 |
| | Ref. [38] | 99.1474 | 99.0414 | 98.7543 | 97.9046 |
| 0.4 | Present | 88.4345 | 90.9806 | 92.8200 | 95.3023 |
| | Ref. [38] | 86.6608 | 86.6414 | 86.4932 | 85.9176 |
| 0.6 | Present | 80.9747 | 84.4598 | 86.8967 | 90.0855 |
| | Ref. [38] | 72.9681 | 73.0382 | 73.0375 | 72.7615 |
| 0.8 | Present | 76.7690 | 80.8426 | 83.5867 | 87.0576 |
| | Ref. [38] | 56.9674 | 57.1341 | 57.2991 | 57.3787 |

**Table 7.**
*Nondimensional natural frequencies of the AFG double-tapered beam (case 4); boundary conditions: C-C.*

$$E(x) = E_L\left(1 - f_E\frac{x}{L}\right), \qquad \rho(x) = \rho_L\left(1 - f_\rho\frac{x}{L}\right) \tag{28}$$

where $f_E = 1 - \frac{E_R}{E_L}$, $\rho_E = 1 - \frac{\rho_R}{\rho_L}$. $E_L$ and $E_R$ are Young's modulus at the left/right end of the beam, and $\rho_L$ and $\rho_R$ are the mass density at the left/right end of the beam. Eq. (2) is then rewritten as

$$\left[(1 - f_E x)w''\right]'' + \left(1 - f_\rho x\right)\ddot{w} = 0 \tag{29}$$

Based on the vibration theory, we assume $w(x,t) = \phi(x)q(t)$, where $q_n(\mathrm{t}) = A_n\cos{\beta_n}^2 t + B_n\sin{\beta_n}^2 t$ and ${\beta_n}^2$ is the modal frequency for dimensionless. The governing equation is then derived as

$$\left[(1-f_E x)\phi_n''\right]'' - \beta_n{}^4\left(1-f_\rho x\right)\phi_n = 0 \tag{30}$$

Next, Meijer G-function will be used to solve the linear partial differential equation. The general expression of Meijer G-function differential equation is written as

$$\left[(-1)^{(p-m-n)} z \prod_{l=1}^{p}\left(z\frac{d}{dz}+1-a_l\right) - \prod_{k=1}^{q}\left(\eta\frac{d}{dz}-b_k\right)\right]G(z) = 0 \tag{31}$$

where $m, n, p$ and $q$ are integers satisfying $0 \le m \le q$, $0 \le \mathrm{n} \le p$, G is the dependent variable also known as the Meijer G-function, z is the independent variable, and $a_l$ and $b_k$ are real numbers.

A definition of the Meijer G-function is given by the following path integral in the complex plane, called the Mellin-Barnes type:

$$G_{p,q}^{m,n}\left\langle \begin{matrix} a_1 \ldots a_n, a_{n+1} \ldots a_p \\ b_1 \ldots b_m, b_{m+1} \ldots b_q \end{matrix} \middle| z \right\rangle = \frac{1}{2\pi i}\int_\tau \frac{\prod_{k=1}^{m}\Gamma(\xi-b_k)\prod_{k=1}^{n}\Gamma(1-a_k+\xi)}{\prod_{k=1}^{p}\Gamma(\xi-a_k)\prod_{k=m+1}^{q}\Gamma(1-b_k+\xi)} z^{-\xi} d\xi \tag{32}$$

where an empty product is interpreted as 1, $0 \le m \le q$, $0 \le \mathrm{n} \le p$, and the parameters are such that none of the poles of $\Gamma(\mathrm{b}_j-\xi)$, $(j = 1 \ldots m)$ coincides with the poles of $\Gamma(1-a_j+\xi)$, $(j = 1 \ldots n)$. Where $i$ is a complex number such that $i^2 = -1$.

A special case of Eq. (31) can be expanded by assuming $n = p = 0$ and $q = 4$. We can get that

$$\begin{aligned} & z^4\frac{d^4G}{dz^4} + \left(6-\sum_{k=1}^{4} b_k\right) z^3 \frac{d^3G}{dz^4} + \left(7 - 3\sum_{n=1}^{4} b_k + \sum_{k,\, l=1}^{4} b_k b_l\right) z^2 \frac{d^2G}{dz^2} \\ & + \left[1-\sum_{k=1}^{4} b_k + \sum_{k,\, l=1}^{4} b_k b_l - (b_1b_2+b_1b_3+b_2b_3)b_4\right] z\frac{dG}{dz} \\ & - \left[(-1)^{-m} z - \prod_{k=1}^{4} b_k\right] G = 0 \end{aligned} \tag{33}$$

where $k \ne l$. Although Eq. (30) is not similar to Eq. (33), the two equations can be similar by introducing some transformations:

$$\phi_n(x) = G(z_n(x)), z_n(x) = \left(\frac{\beta_n}{4f_E}\right)^4 (1-f_E x)^4 \tag{34}$$

Eq. (30) is transformed into

$$\eta_n{}^3 G'''' + 5\eta_n{}^2 G''' + \frac{69}{16}\eta_n G'' + \frac{9}{32}G' - \frac{1-f_\rho x}{1-f_E x}G = 0 \tag{35}$$

Because of the difficulty of solving the differential equation with variable coefficients, we can simplify Eq. (35). Let $1-f_E x = 1-f_\rho x = 1-Fx$; it can be rewritten as

$$\eta_n{}^3 G'''' + 5\eta_n{}^2 G''' + \frac{69}{16}\eta_n G'' + \frac{9}{32}G' - G = 0 \tag{36}$$

| Case | $b_1$ | $b_2$ | $b_3$ | $b_4$ |
|---|---|---|---|---|
| 1 | 1/2 | 1/4 | 0 | 1/4 |
| 2 | 1/4 | 1/4 | 0 | 1/2 |
| 3 | 0 | 1/4 | 1/2 | 1/4 |
| 4 | 1/2 | 0 | 1/4 | 1/4 |
| 5 | 0 | 1/2 | 1/4 | 1/4 |
| 6 | 0 | 1/4 | 1/4 | 1/2 |
| 7 | 1/4 | 0 | 1/4 | 1/2 |
| 8 | 1/4 | 1/2 | 0 | 1/4 |
| 9 | 1/4 | 0 | 1/2 | 1/4 |

**Table 8.**
*Possible case of unknown constant $b_k$ of Meijer G-function equation.*

In order to solve the above equation, the coefficients of the ordinary differential Eqs. (33) and (36) are the same, so we can calculate the corresponding values, as shown in **Table 8.**

One set of data can be selected from **Table 8** and expressed in the form of closed solutions of Meijer G-function:

$$\varphi_{1n}(x) = G_{0,4}^{2,0}\left(\frac{1}{4}, \frac{1}{4}, 0, \frac{1}{2}\middle| \frac{\beta_n{}^4(1-Fx)^4}{256F^4}\right) \tag{37}$$

$$\varphi_{2n}(x) = G_{0,4}^{1,0}\left(\frac{1}{4}, 0, \frac{1}{2}, \frac{1}{4}\middle| -\frac{\beta_n{}^4(1-Fx)^4}{256F^4}\right) \tag{38}$$

$$\varphi_{3n}(x) = G_{0,4}^{1,0}\left(0, \frac{1}{2}, \frac{1}{4}, \frac{1}{4}\middle| -\frac{\beta_n{}^4(1-Fx)^4}{256F^4}\right) \tag{39}$$

$$\varphi_{4n}(x) = G_{0,4}^{1,0}\left(\frac{1}{2}, \frac{1}{4}, \frac{1}{4}, 0\middle| -\frac{\beta_n{}^4(1-Fx)^4}{256F^4}\right) \tag{40}$$

Modal modes of beams:

$$\phi_n(x) = C_{1n}\varphi_{1n}(x) + C_{2n}\varphi_{2n}(x) + C_{3n}\varphi_{3n}(x) + C_{4n}\varphi_{4n}(x), \quad \text{n} = 1, 2, 3, \ldots \tag{41}$$

In order to determine the undetermined coefficients $C_i$ and $\beta_n$, the boundary conditions of beams need to be considered:

1. C-F:

$$\begin{pmatrix} \varphi_{1n}(0) & \varphi_{2n}(0) & \varphi_{3n}(0) & \varphi_{4n}(0) \\ \varphi'_{1n}(0) & \varphi'_{2n}(0) & \varphi'_{3n}(0) & \varphi'_{4n}(0) \\ \varphi''_{1n}(1) & \varphi''_{2n}(1) & \varphi''_{3n}(1) & \varphi''_{4n}(1) \\ \varphi'''_{1n}(1) & \varphi'''_{2n}(1) & \varphi'''_{3n}(1) & \varphi'''_{4n}(1) \end{pmatrix} \begin{Bmatrix} C_{1n} \\ C_{2n} \\ C_{3n} \\ C_{4n} \end{Bmatrix} = \begin{Bmatrix} 0 \\ 0 \\ 0 \\ 0 \end{Bmatrix} \tag{42}$$

1. C-P:

$$\begin{pmatrix} \varphi_{1n}(0) & \varphi_{2n}(0) & \varphi_{3n}(0) & \varphi_{4n}(0) \\ \varphi'_{1n}(0) & \varphi'_{2n}(0) & \varphi'_{3n}(0) & \varphi'_{4n}(0) \\ \varphi_{1n}(1) & \varphi_{2n}(1) & \varphi_{3n}(1) & \varphi_{4n}(1) \\ \varphi''_{1n}(1) & \varphi''_{2n}(1) & \varphi''_{3n}(1) & \varphi''_{4n}(1) \end{pmatrix} \begin{Bmatrix} C_{1n} \\ C_{2n} \\ C_{3n} \\ C_{4n} \end{Bmatrix} = \begin{Bmatrix} 0 \\ 0 \\ 0 \\ 0 \end{Bmatrix} \tag{43}$$

1. S-S:

$$\begin{pmatrix} \varphi_{1n}(0) & \varphi_{2n}(0) & \varphi_{3n}(0) & \varphi_{4n}(0) \\ \varphi''_{1n}(0) & \varphi''_{2n}(0) & \varphi''_{3n}(0) & \varphi''_{4n}(0) \\ \varphi_{1n}(1) & \varphi_{2n}(1) & \varphi_{3n}(1) & \varphi_{4n}(1) \\ \varphi''_{1n}(1) & \varphi''_{2n}(1) & \varphi''_{3n}(1) & \varphi''_{4n}(1) \end{pmatrix} \begin{Bmatrix} C_{1n} \\ C_{2n} \\ C_{3n} \\ C_{4n} \end{Bmatrix} = \begin{Bmatrix} 0 \\ 0 \\ 0 \\ 0 \end{Bmatrix} \tag{44}$$

1. C-C:

$$\begin{pmatrix} \varphi_{1n}(0) & \varphi_{2n}(0) & \varphi_{3n}(0) & \varphi_{4n}(0) \\ \varphi''_{1n}(0) & \varphi''_{2n}(0) & \varphi''_{3n}(0) & \varphi''_{4n}(0) \\ \varphi_{1n}(1) & \varphi_{2n}(1) & \varphi_{3n}(1) & \varphi_{4n}(1) \\ \varphi'_{1n}(1) & \varphi'_{2n}(1) & \varphi'_{3n}(1) & \varphi'_{4n}(1) \end{pmatrix} \begin{Bmatrix} C_{1n} \\ C_{2n} \\ C_{3n} \\ C_{4n} \end{Bmatrix} = \begin{Bmatrix} 0 \\ 0 \\ 0 \\ 0 \end{Bmatrix} \tag{45}$$

### 4.2 Numerical results and discussion

Based on the above analysis, the natural frequencies of beams under different boundary conditions can be solved. Meanwhile, the results of finite element method are also conducted to verify the accuracy of the analytical results. Here, we use the power law gradient of the existing AFG beams [44], and the material properties of AFG beams change continuously along the axial direction. Therefore, the expressions of Young's modulus $E(x)$ and mass density $\rho(x)$ are listed in detail:

$$Y(x) = \begin{cases} Y_L\left(1 - \dfrac{e^{\alpha x/L} - 1}{e^{\alpha} - 1}\right) + Y_R\dfrac{e^{\alpha x/L} - 1}{e^{\alpha} - 1}, & \alpha \neq 0, \\ Y_L\left(1 - \dfrac{x}{L}\right) + Y_R\dfrac{x}{L}, & \alpha = 0. \end{cases} \tag{46}$$

where $Y_L$ and $Y_R$ denote the corresponding material properties of the left and right sides of the beam, respectively. $\alpha$ is the gradient parameter describing the volume fraction change of both constituents involved. When gradient parameter $\alpha$ is equal and less than zero, Young's modulus and mass density at the left end are less than those at the right end. When $\alpha$ equals zero, the beam is equivalent to a uniform Euler-Bernoulli beam, and Young's modulus and mass density of the beam do not change with the length direction of the beam.

The variation of $Y(x)$ along the axis direction of the beam can be shown in **Figure 3** for $Y_R = 3Y_L$. In order to show the practicability of this method, we choose the existing materials to study. The materials of AFG beams are composed of aluminum $(Al)$ and zirconia $(ZrO_2)$.The left and right ends of the beam are pure aluminum and pure zirconia, respectively. The material properties of AFG beams are given in detail in **Table 9**. We choose the sizes of commonly used beams which are $L = 0.2$ m, $B = 0.02$ m, and $H = 0.001$ m.

In order to verify the correctness of this method, some finite element simulation software is used to verify its correctness. In this paper, we analyze the natural frequencies of uniform AFG beams under different boundary conditions. In the process of finite element analysis, the AFG beam is transformed into a finite length model by using the delamination method [85]. At the same time, the AFG beam is delaminated along the axial direction. As shown in **Figure 4**, the material properties change along the axial direction, and the material properties of the adjacent layers are different. In order to analyze the performance of the beam, the uniform element is used to mesh each layer. In order to make the natural frequencies of AFG beams more precise, we can increase the number of layers and refine the finite element meshes.

In the Meijer G-function method, in order to solve the linear natural frequencies of beams under different boundary conditions, the determinant of the coefficient matrix of Eqs. (42)–(45) is equal to zero. Finally, linear natural frequencies of beams with different boundary conditions of the first four orders are listed in **Table 10**.

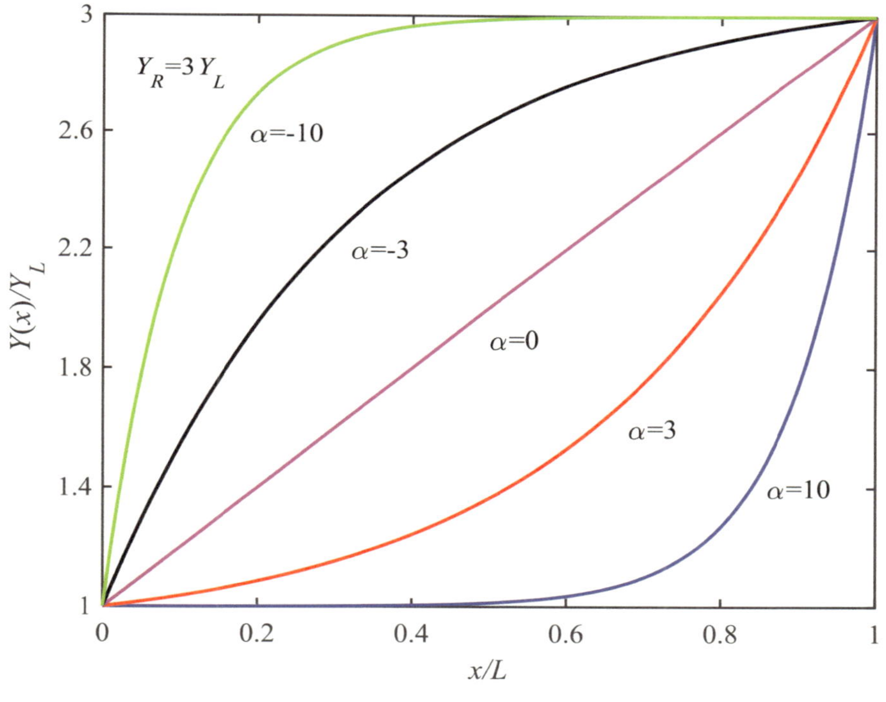


**Figure 3.**
*Variation of the material properties defined by Eq. (46) with $Y_R = 3Y_L$.*

| Properties | Unit | Aluminum | Zirconia |
|---|---|---|---|
| E | GPa | 70 | 200 |
| $\rho$ | Kg/m$^3$ | 2702 | 5700 |

**Table 9.**
*Material properties of the AFG beam [44].*

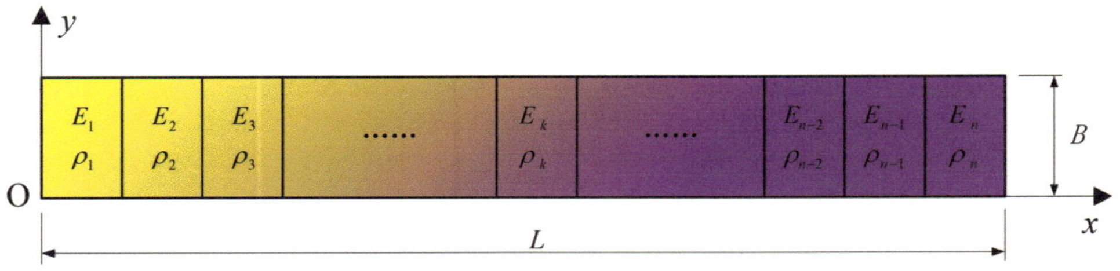


**Figure 4.**
*Finite segment model of the AFG beam.*

| F | Order | C-F | | C-P | | S-S | | C-C | |
|---|---|---|---|---|---|---|---|---|---|
| | | Present | FEM | Present | FEM | Present | FEM | Present | FEM |
| $\alpha = 0.9$ | 1 | 2.4641 | 2.4651 | 4.0787 | 4.0789 | 3.0888 | 3.0891 | 4.5585 | 4.5579 |
| | 2 | 5.2251 | 5.2265 | 7.1520 | 7.1516 | 6.2895 | 6.2893 | 7.6920 | 7.6908 |
| | 3 | 8.2540 | 8.2560 | 10.2762 | 10.2778 | 9.4410 | 9.4420 | 10.8549 | 10.8560 |
| | 4 | 11.3209 | 11.324 | 13.4075 | 13.4092 | 12.5854 | 12.5869 | 14.0136 | 14.0148 |
| $\alpha = 0.5$ | 1 | 2.0774 | 2.0772 | 4.0055 | 4.0056 | 3.1344 | 3.1344 | 4.7098 | 4.7096 |
| | 2 | 4.8497 | 4.8491 | 7.1104 | 7.1100 | 6.2859 | 6.2861 | 7.8364 | 7.8360 |
| | 3 | 7.9496 | 7.9501 | 10.2396 | 10.2409 | 9.4278 | 9.4294 | 10.9827 | 10.9836 |
| | 4 | 11.0455 | 11.0652 | 13.3748 | 13.3760 | 12.5668 | 12.5707 | 14.1256 | 14.1273 |
| $\alpha = 1.7$ | 1 | 1.6098 | 1.6104 | 3.7738 | 3.7738 | 3.1279 | 3.1277 | 4.6896 | 4.6897 |
| | 2 | 4.4786 | 4.4792 | 6.9816 | 6.9816 | 6.2887 | 6.2884 | 7.8189 | 7.8187 |
| | 3 | 7.7326 | 7.7332 | 10.1477 | 10.1490 | 9.4315 | 9.4325 | 10.9679 | 10.9696 |
| | 4 | 10.9067 | 10.9079 | 13.3045 | 13.3042 | 12.5726 | 12.5737 | 14.1139 | 14.1158 |
| $\alpha = 2.7$ | 1 | 1.5370 | 1.5377 | 3.7214 | 3.7217 | 3.1188 | 3.1212 | 4.6629 | 4.6630 |
| | 2 | 4.4142 | 4.4142 | 6.9470 | 6.9471 | 6.2907 | 6.2904 | 7.7947 | 7.7948 |
| | 3 | 7.6920 | 7.6931 | 10.1207 | 10.1222 | 9.4350 | 9.4360 | 10.9482 | 10.9497 |
| | 4 | 10.8759 | 108,772 | 13.2807 | 13.2823 | 12.5762 | 12.5774 | 14.0972 | 14.0988 |

**Table 10.**
*Comparisons between FEM and numerical calculation of linear dimensionless natural frequencies of AFG beams with different boundary conditions.*

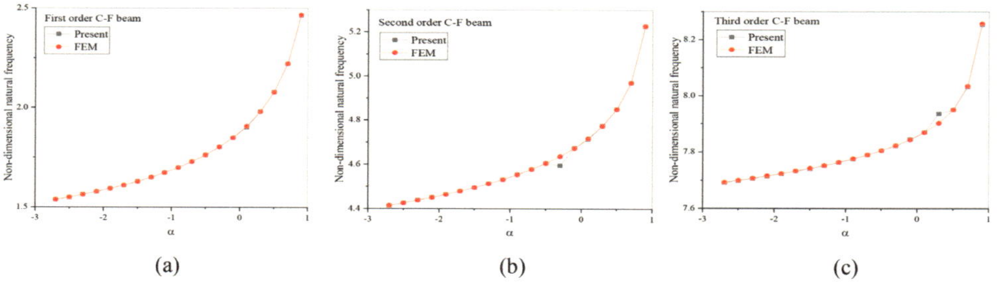


**Figure 5.**
*Dimensionless natural frequencies of C-F beams vary with parameter F: (a) fundamental frequencies, (b) second-order frequencies, and (c) third-order frequencies.*

From **Table 10**, we can see that the results of finite element method are similar to those of Meijer G-function and the error is small. This can prove the accuracy of the method in frequency calculation on the one hand. In **Figure 5**, we can find that the first third-order dimensionless natural frequencies of C-F beams are in good agreement with FEM and numerical calculation. With the gradual increase of gradient parameter F, the dimensionless natural frequency of C-F beam increases gradually, and the change speed is accelerated. At the same time, the FEM and numerical simulation errors are very small, so the precise linear natural frequencies can be obtained.

## 5. Conclusions

FGMs are innovative materials and are very important in engineering and other applications. Despite the variety of methods and approaches for numerical and analytical investigation of nonuniform FG beams, no simple and fast analytical method applicable for such beams with different boundary conditions and varying cross-sectional area was proposed. In this topic, two analytical approaches, the asymptotic perturbation and the Meijer G-function method, were described to analyze the free vibration of the AFG beams.

Based on the Euler-Bernoulli beam theory, the governing differential equations and related boundary conditions are described, which is more complicated because of the partial differential equation with variable coefficients. For both the asymptotic perturbation and the Meijer G-function method, the variable flexural rigidity and mass density are divided into invariant parts and variable parts firstly. Different analytical processes are then carried out to deal with the variable parts applying perturbation theory and the Meijer G-function, respectively. Finally, the simple formulas are derived for solving the nature frequencies of the AFG beams with C-F boundary conditions followed with C-C, C-S, and C-P conditions, respectively. It is observed that natural frequency increases gradually with the increase of the gradient parameter.

Accuracy of the results is also examined using the available data in the published literature and the finite element method. In fact, it can be clearly found that result of the APM is more accurate in low-order mode, which is caused by the defect of the perturbation theory. However, the APM is simple and easily comprehensible, while the Meijer G-function method is more complex and unintelligible for engineers. In general, the results show that the proposed two analytical methods are efficient and can be used to analyze the free vibration of AFG beams.

## Acknowledgements

The authors gratefully acknowledge the support of the National Natural Science Foundation of China (Grant Nos. 11672008, 11702188, and 11272016).

## Author details

Dongxing Cao[1,2*], Bin Wang[1,2], Wenhua Hu[3] and Yanhui Gao[1,2]

1 College of Mechanical Engineering, Beijing University of Technology, Beijing, China

2 Beijing Key Laboratory of Nonlinear Vibrations and Strength of Mechanical Structures, Beijing, China

3 School of Mechanical Engineering, Tianjin University of Technology, Tianjin, China

*Address all correspondence to: caostar@bjut.edu.cn

## References

[1] Nie GJ, Zhong Z, Chen S. Analytical solution for a functionally graded beam with arbitrary graded material properties. Composites Part B: Engineering. 2013;**44**(1):274-282. DOI: 10.1016/j.compositesb.2012.05.029

[2] Nguyen DK. Large displacement response of tapered cantilever beams made of axially functionally graded material. Composites Part B: Engineering. 2013;**55**(9):298-305. DOI: 10.1016/j.compositesb.2013.06.024

[3] Calim FF. Transient analysis of axially functionally graded Timoshenko beams with variable cross-section. Composites Part B: Engineering. 2016; **98**(2015):472-483. DOI: 10.1016/j.compositesb.2016.05.040

[4] Navvab S, Mohammad K, Majid G. Nonlinear vibration of axially functionally graded tapered microbeams. International Journal of Engineering Science. 2016;**102**(2016): 12-26. DOI: 10.1016/j.ijengsci.2016.02.007

[5] Hao YX, Chen LH, Zhang W, Lei JG. Nonlinear oscillations, bifurcations and chaos of functionally graded materials plate. Journal of Sound and Vibration. 2008;**312**(4–5):862-892. DOI: 10.1016/j.jsv.2007.11.033

[6] Hao YX, Zhang W, Yang J, Li S. Nonlinear dynamics of a functionally graded thin simply-supported plate under a hypersonic flow. Mechanics of Advanced Materials and Structures. 2015;**22**(8):619-632. DOI: 10.1080/15376494.2013.828817

[7] Niu Y, Hao Y, Yao M, Zhang W, Yang S. Nonlinear dynamics of imperfect FGM conical panel. Shock and Vibration. 2018;**2018**:1-20. DOI: 10.1155/2018/4187386

[8] Zhang W, Yang J, Hao Y. Chaotic vibrations of an orthotropic FGM rectangular plate based on third-order shear deformation theory. Nonlinear Dynamics. 2010;**59**(4):619-660. DOI: 10.1007/s11071-009-9568-y

[9] Hao YX, Li ZN, Zhang W, Li SB, Yao MH. Vibration of functionally graded sandwich doubly curved shells using improved shear deformation theory. Science China Technological Sciences. 2018;**61**(6):791-808. DOI: 10.1007/s11431-016-9097-7

[10] Hao YX, Zhang W, Yang J. Nonlinear dynamics of cantilever FGM cylindrical shell under 1:2 internal resonance relations. Mechanics of Advanced Materials and Structures. 2012;**20**(10):819-833. DOI: 10.1080/15376494.2012.676717

[11] Zhang W, Hao YX, Yang J. Nonlinear dynamics of FGM circular cylindrical shell with clamped-clamped edges. Composite Structures. 2012; **94**(3):1075-1086. DOI: 10.1016/j.compstruct.2011.11.004

[12] Hao YX, Cao Z, Zhang W, Chen J, Yao MH. Stability analysis for geometric nonlinear functionally graded sandwich shallow shell using a new developed displacement field. Composite Structures. 2019;**210**:202-216. DOI: 10.1016/j.compstruct.2018.11.027

[13] Lee JW, Lee JY. Free vibration analysis of functionally graded Bernoulli-Euler beams using an exact transfer matrix expression. International Journal of Mechanical Sciences. 2017;**122**:1-17. DOI: 10.1016/j.ijmecsci.2017.01.011

[14] Su H, Banerjee JR, Cheung CW. Dynamic stiffness formulation and free vibration analysis of functionally graded beams. Composite Structures. 2013;

**106**(12):854-862. DOI: 10.1016/j.compstruct.2013.06.029

[15] L-l J, P-j M, W-p Z, L-r F, Y-p C. Static and free vibration analysis of functionally graded beams by combination Timoshenko theory and finite volume method. Composite Structures. 2016;**138**:192-213. DOI: 10.1016/j.compstruct.2015.11.027

[16] Ait Atmane H, Tounsi A, Meftah SA, Belhadj HA. Free vibration behavior of exponential functionally graded beams with varying cross-section. Journal of Vibration and Control. 2010; **17**(2):311-318. DOI: 10.1177/1077546310370691

[17] Sina SA, Navazi HM, Haddadpour H. An analytical method for free vibration analysis of functionally graded beams. Materials and Design. 2009; **30**(3):741-747. DOI: 10.1016/j.matdes.2008.05.015

[18] Sharma P. Efficacy of harmonic differential quadrature method to vibration analysis of FGPM beam. Composite Structures. 2018;**189**(2018): 107-116. DOI: 10.1016/j.compstruct.2018.01.059

[19] Li X-F, Wang B-L, Han J-C. A higher-order theory for static and dynamic analyses of functionally graded beams. Archive of Applied Mechanics. 2010;**80**(10):1197-1212. DOI: 10.1007/s00419-010-0435-6

[20] Celebi K, Yarimpabuc D, Tutuncu N. Free vibration analysis of functionally graded beams using complementary functions method. Archive of Applied Mechanics. 2018; **88**(5):729-739. DOI: 10.1007/s00419-017-1338-6

[21] Chen D, Kitipornchai S, Yang J. Nonlinear free vibration of shear deformable sandwich beam with a functionally graded porous core. Thin-Walled Structures. 2016; **107**(2016):39-48. DOI: 10.1016/j.tws.2016.05.025

[22] Nazemnezhad R, Hosseini-Hashemi S. Nonlocal nonlinear free vibration of functionally graded nanobeams. Composite Structures. 2014;**110**: 192-199. DOI: 10.1016/j.compstruct.2013.12.006

[23] Ebrahimi F, Salari E, Hosseini SAH. Thermomechanical vibration behavior of FG nanobeams subjected to linear and non-linear temperature distributions. Journal of Thermal Stresses. 2015; **38**(12):1360-1386. DOI: 10.1080/01495739.2015.1073980

[24] Ebrahimi F, Salari E. Size-dependent thermo-electrical buckling analysis of functionally graded piezoelectric nanobeams. Smart Materials and Structures. 2015;**24**(12): 1-17. DOI:10.1088/0964-1726/24/12/125007

[25] Ebrahimi F, Salari E. Effect of various thermal loadings on buckling and vibrational characteristics of nonlocal temperature-dependent functionally graded nanobeams. Mechanics of Advanced Materials and Structures. 2016;**23**(12):1379-1397. DOI: 10.1080/15376494.2015.1091524

[26] Ebrahimi F, Salari E, Hosseini SAH. In-plane thermal loading effects on vibrational characteristics of functionally graded nanobeams. Meccanica. 2015;**51**(4):951-977. DOI: 10.1007/s11012-015-0248-3

[27] Salari FEaE. A semianalytical method for vibrational and buckling analysis of functionally graded nanobeams considering the physical neutral Axis position. Computer Modeling in Engineering and Sciences. 2015;**105**(2):151-181. DOI: 10.3970/cmes.2015.105.151

[28] Ebrahimi F, Shafiei N. Influence of initial shear stress on the vibration behavior of single-layered graphene sheets embedded in an elastic medium based on Reddy's higher-order shear deformation plate theory. Mechanics of Advanced Materials and Structures. 2016;**24**(9):761-772. DOI: 10.1080/15376494.2016.1196781

[29] Ebrahimi F, Farazmandnia N. Thermo-mechanical vibration analysis of sandwich beams with functionally graded carbon nanotube-reinforced composite face sheets based on a higher-order shear deformation beam theory. Mechanics of Advanced Materials and Structures. 2017;**24**(10):820-829. DOI: 10.1080/15376494.2016.1196786

[30] Ebrahimi F, Salari E. Thermo-mechanical vibration analysis of a single-walled carbon nanotube embedded in an elastic medium based on higher-order shear deformation beam theory. Journal of Mechanical Science and Technology. 2015;**29**(9): 3797-3803. DOI: 10.1007/s12206-015-0826-2

[31] Ebrahimi F, Ghasemi F, Salari E. Investigating thermal effects on vibration behavior of temperature-dependent compositionally graded Euler beams with porosities. Meccanica. 2015; **51**(1):223-249. DOI: 10.1007/s11012-015-0208-y

[32] Ebrahimi F, Mokhtari M. Transverse vibration analysis of rotating porous beam with functionally graded microstructure using the differential transform method. Journal of The Brazilian Society of Mechanical Sciences and Engineering. 2015;**37**(4):1435-1444

[33] Ebrahimi F, Zia M. Large amplitude nonlinear vibration analysis of functionally graded Timoshenko beams with porosities. Acta Astronautica. 2015; **116**:117-125. DOI: 10.1016/j.actaastro.2015.06.014

[34] Ebrahimi F, Salari E. Thermal buckling and free vibration analysis of size dependent Timoshenko FG nanobeams in thermal environments. Composite Structures. 2015;**128**: 363-380. DOI: 10.1016/j.compstruct.2015.03.023

[35] Ebrahimi F, Salari E. Thermo-mechanical vibration analysis of nonlocal temperature-dependent FG nanobeams with various boundary conditions. Composites Part B: Engineering. 2015;**78**:272-290. DOI: 10.1016/j.compositesb.2015.03.068

[36] Ebrahimi F, Salari E. Nonlocal thermo-mechanical vibration analysis of functionally graded nanobeams in thermal environment. Acta Astronautica. 2015;**113**:29-50

[37] Alshorbagy AE, Eltaher MA, Mahmoud FF. Free vibration characteristics of a functionally graded beam by finite element method. Applied Mathematical Modelling. 2011;**35**(1): 412-425. DOI: 10.1016/j.apm.2010.07.006

[38] Shahba A, Attarnejad R, Hajilar S. Free vibration and stability of axially functionally graded tapered Euler-Bernoulli beams. Shock and Vibration. 2011;**18**(5):683-696. DOI: 10.3233/sav-2010-0589

[39] Shahba A, Attarnejad R, Marvi MT, Hajilar S. Free vibration and stability analysis of axially functionally graded tapered Timoshenko beams with classical and non-classical boundary conditions. Composites Part B: Engineering. 2011;**42**(4):801-808. DOI: 10.1016/j.compositesb.2011.01.017

[40] Shahba A, Rajasekaran S. Free vibration and stability of tapered Euler-Bernoulli beams made of axially functionally graded materials. Applied Mathematical Modelling. 2012;**36**(7): 3094-3111. DOI: 10.1016/j.apm.2011.09.073

[41] Liu P, Lin K, Liu H, Qin R. Free transverse vibration analysis of axially functionally graded tapered Euler-Bernoulli beams through spline finite point method. Shock and Vibration. 2016;**2016**(5891030):1-23. DOI: 10.1155/2016/5891030

[42] Rajasekaran S. Differential transformation and differential quadrature methods for centrifugally stiffened axially functionally graded tapered beams. International Journal of Mechanical Sciences. 2013;**74**:15-31. DOI: 10.1016/j.ijmecsci.2013.04.004

[43] Rajasekaran S, Tochaei EN. Free vibration analysis of axially functionally graded tapered Timoshenko beams using differential transformation element method and differential quadrature element method of lowest-order. Meccanica. 2014;**49**(4):995-1009. DOI: 10.1007/s11012-013-9847-z

[44] Huang Y, Li X-F. A new approach for free vibration of axially functionally graded beams with non-uniform cross-section. Journal of Sound and Vibration. 2010;**329**(11):2291-2303. DOI: 10.1016/j.jsv.2009.12.029

[45] Huang Y, Yang L-E, Luo Q-Z. Free vibration of axially functionally graded Timoshenko beams with non-uniform cross-section. Composites Part B: Engineering. 2013;**45**(1):1493-1498. DOI: 10.1016/j.compositesb.2012.09.015

[46] Huang Y, Rong H-W. Free vibration of axially inhomogeneous beams that are made of functionally graded materials. The International Journal of Acoustics and Vibration. 2017;**22**(1):68-73. DOI: 10.20855/ijav.2017.22.1452

[47] Hein H, Feklistova L. Free vibrations of non-uniform and axially functionally graded beams using Haar wavelets. Engineering Structures. 2011;**33**(12):3696-3701. DOI: 10.1016/j.engstruct.2011.08.006

[48] Xie X, Zheng H, Zou X. An integrated spectral collocation approach for the static and free vibration analyses of axially functionally graded nonuniform beams. Proceedings of the Institution of Mechanical Engineers, Part C: Journal of Mechanical Engineering Science. 2016;**231**(13):2459-2471. DOI: 10.1177/0954406216634393

[49] Kukla S, Rychlewska J. An approach for free vibration analysis of axially graded beams. Journal of Theoretical and Applied Mechanics. 2016;**54**(3):859-870. DOI: 10.15632/jtam-pl.54.3.859

[50] Zhao Y, Huang Y, Guo M. A novel approach for free vibration of axially functionally graded beams with non-uniform cross-section based on Chebyshev polynomials theory. Composite Structures. 2017;**168**:277-284. DOI: 10.1016/j.compstruct.2017.02.012

[51] Fang J, Zhou D. Free vibration analysis of rotating axially functionally graded-tapered beams using Chebyshev-Ritz method. Materials Research Innovations. 2015;**19**(5):S5-1255-S5S5-62. DOI: 10.1179/1432891714Z.0000000001289

[52] Fang JS, Zhou D. Free vibration analysis of rotating axially functionally graded tapered Timoshenko beams. International Journal of Structural Stability and Dynamics. 2016;**16**(05):1-19. DOI: 10.1142/S0219455415500078

[53] Li XF, Kang YA, Wu JX. Exact frequency equations of free vibration of exponentially functionally graded beams. Applied Acoustics. 2013;**74**(3):413-420. DOI: 10.1016/j.apacoust.2012.08.003

[54] Tang AY, Wu JX, Li XF, Lee KY. Exact frequency equations of free vibration of exponentially non-uniform functionally graded Timoshenko beams. International Journal of Mechanical

Sciences. 2014;**89**:1-11. DOI: 10.1016/j.ijmecsci.2014.08.017

[55] Sarkar K, Ganguli R. Closed-form solutions for axially functionally graded Timoshenko beams having uniform cross-section and fixed–fixed boundary condition. Composites Part B: Engineering. 2014;**58**:361-370. DOI: 10.1016/j.compositesb.2013.10.077

[56] Akgöz B, Civalek Ö. Free vibration analysis of axially functionally graded tapered Bernoulli-Euler microbeams based on the modified couple stress theory. Composite Structures. 2013; **98**(3):314-322. DOI: 10.1016/j.compstruct.2012.11.020

[57] Yuan J, Pao Y-H, Chen W. Exact solutions for free vibrations of axially inhomogeneous Timoshenko beams with variable cross section. Acta Mech. 2016;**227**(9):2625-2643. DOI: 10.1007/s00707-016-1658-6

[58] Yilmaz Y, Evran S. Free vibration analysis of axially layered functionally graded short beams using experimental and finite element methods. Science and Engineering of Composite Materials. 2016;**23**(4):453-460. DOI: 10.1515/secm-2014-0161

[59] Boiangiu M, Ceausu V, Untaroiu CD. A transfer matrix method for free vibration analysis of Euler-Bernoulli beams with variable cross section. Journal of Vibration and Control. 2016; **22**(11):2591-2602. DOI: 10.1177/1077546314550699

[60] Garijo D. Free vibration analysis of non-uniform Euler-Bernoulli beams by means of Bernstein pseudospectral collocation. Engineering with Computers. 2015;**31**(4):813-823. DOI: 10.1007/s00366-015-0401-6

[61] Arndt M, Machado RD, Scremin A. Accurate assessment of natural frequencies for uniform and non-uniform Euler-Bernoulli beams and frames by adaptive generalized finite element method. Engineering Computations. 2016;**33**(5):1586-1609. DOI: 10.1108/EC-05-2015-0116

[62] Zhernakov VS, Pavlov VP, Kudoyarova VM. Spline-method for numerical calculation of natural-vibration frequency of beam with variable cross-section. Procedia Engineering. 2017;**206**:710-715. DOI: 10.1016/j.proeng.2017.10.542

[63] Wang CY. Vibration of a tapered cantilever of constant thickness and linearly tapered width. Archive of Applied Mechanics. 2013;**83**(1):171-176. DOI: 10.1007/s00419-012-0637-1

[64] Silva CJ, Daqaq MF. Nonlinear flexural response of a slender cantilever beam of constant thickness and linearly-varying width to a primary resonance excitation. Journal of Sound and Vibration. 2017;**389**(2017):438-453. DOI: 10.1016/j.jsv.2016.11.029

[65] Rajasekaran S, Khaniki HB. Bending, buckling and vibration of small-scale tapered beams. International Journal of Engineering Science. 2017; **120**:172-188. DOI: 10.1016/j.ijengsci.2017.08.005

[66] Çalım FF. Free and forced vibrations of non-uniform composite beams. Composite Structures. 2009; **88**(3):413-423. DOI: 10.1016/j.compstruct.2008.05.001

[67] Yang X, Wang S, Zhang W, Qin Z, Yang T. Dynamic analysis of a rotating tapered cantilever Timoshenko beam based on the power series method. Applied Mathematics and Mechanics. 2017;**38**(10):1425-1438. DOI: 10.1007/s10483-017-2249-6

[68] Clementi F, Demeio L, Mazzilli CEN, Lenci S. Nonlinear vibrations of non-uniform beams by the MTS asymptotic expansion method.

Continuum Mechanics and Thermodynamics. 2015;**27**(4):703-717. DOI: 10.1007/s00161-014-0368-3

[69] Wang X, Wang Y. Free vibration analysis of multiple-stepped beams by the differential quadrature element method. Applied Mathematics and Computation. 2013;**219**(11):5802-5810. DOI: 10.1016/j.amc.2012.12.037

[70] Abdelghany SM, Ewis KM, Mahmoud AA, Nassar MM. Vibration of a circular beam with variable cross sections using differential transformation method. Beni-Suef University Journal of Basic and Applied Sciences. 2015;**4**(3):185-191. DOI: 10.1016/j.bjbas.2015.05.006

[71] Chen L-Q, Chen H. Asymptotic analysis on nonlinear vibration of axially accelerating viscoelastic strings with the standard linear solid model. Journal of Engineering Mathematics. 2010;**67**(3): 205-218. DOI: 10.1007/s10665-009-9316-9

[72] Yan Q, Ding H, Chen L. Nonlinear dynamics of axially moving viscoelastic Timoshenko beam under parametric and external excitations. Applied Mathematics and Mechanics. 2015; **36**(8):971-984. DOI: 10.1007/s10483-015-1966-7

[73] Ding H, Tang YQ, Chen LQ. Frequencies of transverse vibration of an axially moving viscoelastic beam. Journal of Vibration and Control. 2015; **23**(20):1-11. DOI: 10.1177/1077546315600311

[74] Ding H, Huang L, Mao X, Chen L. Primary resonance of traveling viscoelastic beam under internal resonance. Applied Mathematics and Mechanics. 2016;**38**(1):1-14. DOI: 10.1007/s10483-016-2152-6

[75] Chen RM. Some nonlinear dispersive waves arising in compressible hyperelastic plates. International Journal of Engineering Science. 2006;**44** (18–19):1188-1204. DOI: 10.1016/j.ijengsci.2006.08.003

[76] Hao YX, Zhang W, Yang J. Nonlinear oscillation of a cantilever FGM rectangular plate based on third-order plate theory and asymptotic perturbation method. Composites Part B: Engineering. 2011;**42**(3):402-413. DOI: 10.1016/j.compositesb.2010.12.010

[77] Andrianov IV, Danishevs'Kyy VV. Asymptotic approach for non-linear periodical vibrations of continuous structures. Journal of Sound and Vibration. 2002;**249**(3):465-481. DOI: 10.1006/jsvi.2001.3878

[78] Nayfeh AH, Mook DT. Nonlinear Oscillations. New York: John Wiley and Sons; 1979

[79] Lenci S, Clementi F, Mazzilli CEN. Simple formulas for the natural frequencies of non-uniform cables and beams. International Journal of Mechanical Sciences. 2013;**77**(4): 155-163. DOI: 10.1016/j.ijmecsci.2013.09.028

[80] Cao DX, Gao YH, Wang JJ, Yao MH, Zhang W. Analytical analysis of free vibration of non-uniform and non-homogenous beams: Asymptotic perturbation approach. Applied Mathematical Modelling. 2019;**65**: 526-534. DOI: 10.1016/j.apm.2018.08.026

[81] Cao D, Gao Y. Free vibration of non-uniform axially functionally graded beams using the asymptotic development method. Applied Mathematics and Mechanics (English Edition). 2018;**40**(1):85-96. DOI: 10.1007/s10483-019-2402-9

[82] Cao DX, Gao YH, Yao MH, Zhang W. Free vibration of axially functionally graded beams using the asymptotic development method. Engineering

Structures. 2018;**173**:442-448. DOI: 10.1016/j.engstruct.2018.06.111

[83] Tarnopolskaya T, de Hoog F, Fletcher NH, Thwaites S. Asymptotic analysis of the free in-plane vibrations of beams with arbitrarily varying curvature and cross-section. Journal of Sound and Vibration. 1996;**196**(5): 659-680. DOI: 10.1006/jsvi.1996.0507

[84] Kryzhevich SG, Volpert VA. Different types of solvability conditions for differential operators. Electronic Journal of Differential Equations. 2006; **2006**(100):1-24. DOI: 10.1142/9789812772992_0015

[85] Wang B, Han J, Du S. Dynamic response for functionally graded materials with penny-shaped cracks. Acta Mechanica Solida Sinica. 1999; **12**(2):106-113

Chapter 2

# Boundary Element Model for Nonlinear Fractional-Order Heat Transfer in Magneto-Thermoelastic FGA Structures Involving Three Temperatures

*Mohamed Abdelsabour Fahmy*

## Abstract

The principal objective of this chapter is to introduce a new fractional-order theory for functionally graded anisotropic (FGA) structures. This theory called nonlinear uncoupled magneto-thermoelasticity theory involving three temperatures. Because of strong nonlinearity, it is very difficult to solve the problems related to this theory analytically. Therefore, it is necessary to develop new numerical methods for solving such problems. So, we propose a new boundary element model for the solution of general and complex problems associated with this theory. The numerical results are presented graphically in order to display the effect of the graded parameter on the temperatures and displacements. The numerical results also confirm the validity and accuracy of our proposed model.

**Keywords:** boundary element method, fractional-order heat transfer, functionally graded anisotropic structures, nonlinear uncoupled magneto-thermoelasticity, three temperatures

## 1. Introduction

Functionally graded material (FGM) is a special type of advanced inhomogeneous materials. Functionally graded structure is a mixture of two or more distinct materials (usually heat-resisting ceramic on the outside surface and fracture-resisting metal on the inside surface) that have specified properties in specified direction of the structure to achieve a require function [1, 2]. This feature enables obtaining structures with the best of both material's properties, and suitable for applications requiring high thermal resistance and high mechanical strength [3–12].

Functionally Graded Materials have been wide range of thermoelastic applications in several fields, for example, the water-cooling model of a fusion reactor divertor is one of the most widely used models in industrial design, which is consisting of a tungsten (W) and a copper (Cu), that subjected to a structural integrity issue due to thermal stresses resulted from thermal expansion mismatch between the bond materials. Recently, functionally graded tungsten (W)–copper (Cu) has been developed by using a precipitation-hardened copper alloy as matrix



IntechOpen

instead of pure copper, to overcome the loss of strength due to the softening of the copper matrix.

The carbon nanotubes (CNT) in FGM have new applications such as reinforced functionally graded piezoelectric actuators, reinforced functionally graded polyestercalcium phosphate materials for bone replacement, reinforced functionally graded tools and dies for reduce scrap, better wear resistance, better thermal management, and improved process productivity, reinforced metal matrix functionally graded composites used in mining, geothermal drilling, cutting tools, drills and machining of wear resistant materials. Also, they used as furnace liners and thermal shielding elements in microelectronics.

There are many areas of application for elastic and thermoelastic functionally graded materials, for example, industrial applications such as MRI scanner cryogenic tubes, eyeglass frames, musical instruments, pressure vessels, fuel tanks, cutting tool inserts, laptop cases, wind turbine blades, firefighting air bottles, drilling motor shaft, X-ray tables, helmets and aircraft structures. Automobiles applications such as combustion chambers, engine cylinder liners, leaf springs, diesel engine pistons, shock absorbers, flywheels, drive shafts and racing car brakes. Aerospace applications rocket nozzle, heat exchange panels, spacecraft truss structure, reflectors, solar panels, camera housing, turbine wheels and Space shuttle. Submarine applications such as propulsion shaft, cylindrical pressure hull, sonar domes, diving cylinders and composite piping system. Biotechnology applications such as functional gradient nanohydroxyapatite reinforced polyvinyl alcohol gel biocomposites. Defense applications such as armor plates and bullet-proof vests. High-temperature environment applications such as aerospace and space vehicles. Biomedical applications such as orthopedic applications for teeth and bone replacement. Energy applications such as energy conversion devices and as thermoelectric converter for energy conservation. They also provide thermal barrier and are used as protective coating on turbine blades in gas turbine engine. Marine applications such as parallelogram slabs in buildings and bridges, swept wings of aircrafts and ship hulls. Optoelectronic applications such as automobile engine components, cutting tool insert coating, nuclear reactor components, turbine blade, tribology, sensors, heat exchanger, fire retardant doors, etc.

According to continuous and smooth variation of FGM properties throughout in depth, there are many laws to describe the behavior of FGM such as index [13], sigmoid law [14], exponential law [15] and power law [16–24].

There was widespread interest in functionally graded materials, which has developed a lot of analytical methods for analysis of elasticity [25–32] and thermoelasticity [33–53] problems, some of which have become dominant in scientific literature. For the numerical methods, the isogeometric finite element method (FEM) has been used by Valizadeh et al. [54] for static characteristics of FGM and by Bhardwaj et al. [55] for solving crack problem of FGM. Nowadays, the boundary element method is a simple, efficient and powerful numerical tool which provides an excellent alternative to the finite element method for the solution of FGM problems, Sladek et al. [56–58] have been developed BEM formulation for transient thermal problems in FGMs. Gao et al. [59] developed fracture analysis of functionally graded materials by a BEM. Fahmy [60–72] developed BEM to solve elastic, thermoelastic and biomechanic problems in anisotropic functionally graded structures. Further details on the BEM are given in [73, 74] and the references therein.

In the present paper, we propose new FGA structures theory and new boundary element technique for modeling problems of nonlinear uncoupled magneto-thermoelasticity involving three temperatures. The boundary element method reduces the dimension of the problem, therefore, we obtain a reduction of numerical approximation, linear equations system and input data. Since there is strong nonlinearity in the proposed theory and its related problems. So, we develop new boundary element technique for modeling such problems. The numerical results are presented graphically through the thickness of the homogeneous and functionally graded structures to show the effect of graded parameter on the temperatures and displacements. The numerical results demonstrate the validity and accuracy of our proposed model.

A brief summary of the chapter is as follows: Section 1 outlines the background and provides the readers with the necessary information to books and articles for a better understanding of mechanical behaviour of magneto-thermoelastic FGA structures and their applications. Section 2 describes the formulation of the new theory and its related problems. Section 3 discusses the implementation of the new BEM for solving the nonlinear radiative heat conduction equation, to obtain the three temperature fields. Section 4 studies the development of new BEM and its implementation for solving the move equation based on the known three temperature fields, to obtain the displacement field. Section 5 presents the new numerical results that describe the through-thickness mechanical behaviour of homogeneous and functionally graded structures.

## 2. Formulation of the problem

We consider a Cartesian coordinate system for 2D structure (see **Figure 1**) which is functionally graded along the 0x direction, and considering z-axis is the direction of the effect of the constant magnetic field $H_0$ .

The fractional-order governing equations of three temperatures nonlinear uncoupled magneto-thermoelasticity in FGA structures can be written as follows [6].

$$\sigma_{pj,j} + \tau_{pj,j} = \rho(x+1)^m \ddot{u}_k \tag{1}$$

$$\sigma_{pj} = (x+1)^m \left[C_{pjkl} u_{k,l} \text{-} \beta_{pj} T_\alpha(r,\tau)\right] \tag{2}$$

$$\tau_{pj} = \mu(x+1)^m \left(h_p H_j + H_j h_p \text{-} \delta_{pj}(h_k H_k)\right) \tag{3}$$

where $\sigma_{pj}$,$\tau_{pj}$, $u_k$, $C_{pjkl}$ ($C_{pjkl} = C_{klpj} = C_{kljp}$), $\beta_{pj}$ ($\beta_{pj} = \beta_{jp}$), $\mu$ and $h_p$ are respectively, mechanic stress tensor, Maxwell stress tensor, displacement, constant elastic moduli, stress-temperature coefficients, magnetic permeability and perturbed magnetic field.

The nonlinear time-dependent two dimensions three temperature (2D-3 T) radiation diffusion equations coupled by electron, ion and phonon temperatures may be written as follows

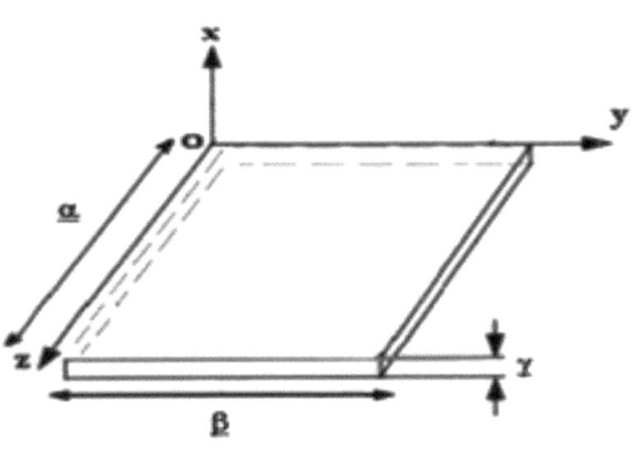


**Figure 1.**
*Geometry of the FGA structure.*

$$D_{\tau}^{a}T_{\alpha}(r,\tau)=\xi\nabla[\mathbb{K}_{\alpha}\,\nabla T_{\alpha}(r,\tau)]+\xi\overline{\mathbb{W}}(r,\tau),\xi=\frac{1}{c_{\alpha}\rho\delta_{1}} \tag{4}$$

where

$$\overline{\mathbb{W}}(x,y,\tau)=\begin{cases}-\rho\,\mathbb{W}_{ei}\,(T_e\text{-}T_i)-\rho\,\mathbb{W}_{er}\,(T_e\text{-}T_r), & \alpha=e,\delta_1=1\\ \rho\,\mathbb{W}_{ei}\,(T_e\text{-}T_i), & \alpha=i,\ \delta_1=1\\ \rho\,\mathbb{W}_{ep}\,(T_e\text{-}T_p), & \alpha=p,\delta_1=\dfrac{4}{\rho}T_p^3\end{cases} \tag{5}$$

and

$$\mathbb{W}_{ei}=\rho\mathbb{A}_{ei}T_e^{-2/3},\ \mathbb{W}_{ep}=\rho\mathbb{A}_{ep}T_e^{-1/2}\ \mathbb{K}_{\alpha}=\mathbb{A}_{\alpha}T_{\alpha}^{5/2},\alpha=e,i,\ \mathbb{K}_p=\mathbb{A}_pT_p^{3+\mathbb{B}} \tag{6}$$

The total energy per unit mass can be expressed as follows

$$P=P_e+P_i+P_p,P_e=c_eT_e,P_i=c_iT_i,P_p=\frac{1}{\rho}c_pT_p^4 \tag{7}$$

where $\mathbb{K}_{\alpha}$ are conductive coefficients, $T_{\alpha}$ are temperature functions, $c_{\alpha}(\alpha=e,i,p)$ are isochore specific-heat coefficients, $\rho$ is the density, $\tau$ is the time. In which, $c_{\alpha}$, $\mathbb{A}_{\alpha}$ $(\alpha=e,i,p)$, $\mathbb{B}$, $\mathbb{A}_{ei}$, $\mathbb{A}_{ep}$ are constant inside each subdomain, $\mathbb{W}_{ei}$ and $\mathbb{W}_{ep}$ are electron-ion coefficient and electron–phonon coefficient, respectively.

Initial and boundary conditions can be written as

$$T_{\alpha}(r,0)=T_{\alpha}^{0}(r)=g_1(r,\tau), \tag{8}$$

$$u_k(r,0)=\dot{u}_k(r,0)=0\ \text{for}\ r\in R\cup C(r,\tau), \tag{9}$$

$$\mathbb{K}_{\alpha}\frac{\partial T_{\alpha}}{\partial n}\bigg|_{\Gamma_1}=0,\alpha=e,i,T_r|_{\Gamma_1}=g_2(r,\tau) \tag{10}$$

$$u_k(r,\tau)=\Psi_k(r,\tau)\ \text{for}\ \ r\in C_3 \tag{11}$$

$$t_k(r,\tau)=\delta_k(r,\tau)\ \text{for}\ \ r\in C_4,C=C_3\cup C_4,C_3\cap C_4=\emptyset \tag{12}$$

$$\mathbb{K}_{\alpha}\frac{\partial T_{\alpha}}{\partial n}\bigg|_{\Gamma_2}=0,\alpha=e,i,p \tag{13}$$

$$T(r,\tau)=\mathrm{H}(r,\tau)\ \text{for}\ \ r\in C_1,\qquad \tau>0 \tag{14}$$

$$q(r,\tau)=h(r,\tau)\ \text{for}\ \ r\in C_2,\qquad \tau>0\,,C=C_1\cup C_2,C_1\cap C_2=\emptyset \tag{15}$$

## 3. BEM numerical implementation for temperature field

This section outlines the solution of 2D nonlinear time-dependent three temperatures (electron, ion and phonon) radiation diffusion equations using a boundary element method.

Now, let us consider $\xi = \frac{1}{c_\alpha \rho \delta_1} = 1$ and discretize the time interval $[0, F]$ into $F + 1$ equal time steps, where $\tau^f = f\Delta\tau, f = 0, 1, 2, \ldots, F$ Let $T^f(\boldsymbol{x}) = T(\boldsymbol{x}, \tau^f)$ be the solution at time $\tau^f$. Assuming that the time derivative of temperature within the time interval $[\tau^f, \tau^{f+1}]$ can be approximated by.

$$\dot{T}_\alpha(r,\tau) = \frac{T_\alpha^{f+1}(r) - T_\alpha^f(r)}{\Delta\tau} + O(\Delta\tau) \tag{16}$$

$\boldsymbol{D_\tau^a}$ denotes the Caputo fractional time derivative of order $a$ defined by [75].

$$D_\tau^a T_\alpha(r,\tau) = \frac{1}{\Gamma(1-a)} \int_0^\tau \frac{\partial T_\alpha(r,s)}{\partial s} \frac{ds}{(\tau-s)^a}, 0 < a < 1 \tag{17}$$

By using a finite difference scheme of Caputo fractional time derivative of order $a$ (17) at times $(f+1)\Delta\tau$ and $f\Delta\tau$, we obtain:

$$D_\tau^a T_\alpha^{f+1} + D_\tau^a T_\alpha^f \approx \sum_{j=0}^{k} W_{a,j} \left( T_\alpha^{f+1-j}(r) - T_\alpha^{f-j}(r) \right), (f = 1, 2, \ldots, F) \tag{18}$$

Where

$$W_{a,0} = \frac{(\Delta\tau)^{-a}}{\Gamma(2-a)} \tag{19}$$

$$W_{a,j} = W_{a,0}\left((j+1)^{1-a} - (j-1)^{1-a}\right), j = 1, 2, \ldots, F \tag{20}$$

According to Eq. (18), the fractional order heat Eq. (4) can be replaced by the following system

$$\begin{aligned} W_{a,0} T_\alpha^{f+1}(r) - \mathbb{K}_\alpha(x) T_{\alpha,II}^{f+1}(r) - \mathbb{K}_{\alpha,I}(x) T_{\alpha,I}^{f+1}(r) \\ = W_{a,0} T_\alpha^f(r) - \mathbb{K}_\alpha(x) T_{\alpha,II}^f(r) - \mathbb{K}_{\alpha,I}(x) T_{\alpha,I}^f(r) \\ - \sum_{j=1}^{f} W_{a,j}\left(T_\alpha^{f+1-j}(r) - T_\alpha^{f-j}(r)\right) + \overline{\mathbb{W}}_m^{f+1}(x,\tau) + \overline{\mathbb{W}}_m^f(x,\tau), f = 0, 1, 2, \ldots, F \end{aligned} \tag{21}$$

According to Fahmy [60], and using the fundamental solution which satisfies the system (21), the boundary integral equations corresponding to nonlinear three temperature heat conduction-radiation equations can be written as

$$CT_\alpha = \frac{D}{\mathbb{K}_\alpha} \int_0^\tau \int_S \left[T_\alpha q^* - T_\alpha^* q\right] dS\, d\tau + \frac{D}{\mathbb{K}_\alpha} \int_0^\tau \int_R b\, T_\alpha^*\, dR\, d\tau + \int_R T_\alpha^i T_\alpha^* \big|_{\tau=0} dR \tag{22}$$

which can be written in the absence of internal heat sources as follows

$$CT_\alpha = \int_S [T_\alpha q^* \text{-} T_\alpha^* q]\, dS \text{ - } \int_R \frac{\mathbb{K}_\alpha}{D} \frac{\partial T_\alpha^*}{\partial \tau} T_\alpha \, dR \tag{23}$$

Time temperature derivative can be written as

$$\frac{\partial T_\alpha}{\partial \tau} \cong \sum_{j=1}^{N} f^j(r)^j a^j(\tau) \tag{24}$$

where $f^j(r)$ are known functions and $a^j(\tau)$ are unknown coefficients. We suppose that $\hat{T}_\alpha^j$ is a solution of

$$\nabla^2 \hat{T}_\alpha^j = f^j \tag{25}$$

Then, Eq. (23) yields the following boundary integral equation

$$C\,T = \int_S [T_\alpha q^* \text{-} T_\alpha^* q]\, dS + \sum_{j=1}^{N} a^j(\tau) D^{\text{-}1} \left( C\hat{T}_\alpha^j \text{-} \int_S [T_\alpha^j q^* \text{-} \hat{q}^j T_\alpha^*]\, dS \right) \tag{26}$$

where

$$\hat{q}^j = \text{-}\mathbb{K}_\alpha \frac{\partial \hat{T}_\alpha^j}{\partial n} \tag{27}$$

and

$$a^j(\tau) = \sum_{i=1}^{N} f_{ji}^{\text{-}1} \frac{\partial T(r_i, \tau)}{\partial \tau} \tag{28}$$

In which the entries of $f_{ji}^{-1}$ are the coefficients of $F^{-1}$ with matrix $F$ defined as [76].

$$\{F\}_{ji} = f^j(r_i) \tag{29}$$

Using the standard boundary element discretization scheme for Eq. (26) and using Eq. (28), we have

$$C\,\dot{T}_\alpha + H\,T_\alpha = G\,Q \tag{30}$$

The diffusion matrix can be defined as

$$C = \text{-}[H\,\hat{T}_\alpha \text{-} G\,\hat{Q}] F^{\text{-}1} D^{\text{-}1} \tag{31}$$

with

$$\{\hat{T}\}_{ij} = \hat{T}^j(x_i) \tag{32}$$

$$\{\hat{Q}\}_{ij} = \hat{q}^j(x_i) \tag{33}$$

In order to solve Eq. (30) numerically the functions $T_\alpha$ and $q$ are interpolated as

$$T_\alpha = (1-\theta)\, T_\alpha^m + \theta\, T_\alpha^{m+1} \tag{34}$$

$$q = (1-\theta) q^m + \theta\, q^{m+1} \tag{35}$$

where $\theta = \frac{\tau\text{-}\tau^m}{\tau^{m+1}\text{-}\tau^m}$ , $0 \le \theta \le 1$ determines the practical time $\tau$ in the current time step.

By differentiating Eq. (34) with respect to time we get

$$\dot{T}_\alpha = \frac{dT_\alpha}{d\theta}\frac{d\theta}{d\tau} = \frac{T_\alpha^{m+1}\text{-}T_\alpha^m}{\tau^{m+1}\text{-}\tau^m} = \frac{T_\alpha^{m+1}\text{-}T_\alpha^m}{\Delta\tau^m} \tag{36}$$

The substitution of Eqs. (34)–(36) into Eq. (30) leads to

$$\left(\frac{C}{\Delta\tau^m} + \theta H\right) T_\alpha^{m+1}\text{-}\theta G Q^{m+1} = \left(\frac{C}{\Delta\tau^m}\text{-}(1-\theta)H\right) T_\alpha^m + (1-\theta) G Q^m \tag{37}$$

By using initial and boundary conditions, we get

$$\mathbb{a}X = \mathbb{b} \tag{38}$$

This system yields the temperature, that can be used to solve (1) for the displacement.

## 4. BEM numerical implementation for displacement field

Based on Eqs. (2) and (3), Eq. (1) can be rewritten as

$$L_{jl} u_k = \rho \ddot{u}_p\text{-}\left(D_{jl} u_k + D_{pj} T_\alpha\right) = \rho \ddot{u}_p\text{-}\rho b_p = \rho b_p' \tag{39}$$

where

$$L_{jl} = D_l \frac{\partial}{\partial x_j},\ D_{jl} = \mu H_0^2 D_j \frac{\partial}{\partial x_l},\ D_{pj} = \text{-}\beta_{pj} D_j,\ D_l = C_{pjkl}\frac{\partial}{\partial x_l},\ D_j = \left(\frac{\partial}{\partial x_j} + \Lambda\right),\ \Lambda = \frac{m}{x+1} \tag{40}$$

when the temperatures are known, the displacement can be computed by solving (39) using BEM. By choosing $u_p^*$ as the weight function and applying the weighted residual method, Eq. (39) can be reexpressed as

$$\int_{R} (L_{jl}u_k - \rho b_p^{'})u_p^* dR = 0 \tag{41}$$

The first term in (41) can be integrated partially using Gau β theory yields

$$\int_{R} C_{pjkl}u_{k,lj}u_p^* \, dR = \int_{C} C_{pjkl}u_{k,l}u_p^* n_j dC - \int_{R} C_{pjkl}u_{k,l}u_{p,j}^* dR \tag{42}$$

The last term in (42) can be integrated partially twice using Gau β theory yields

$$\int_{R} C_{pjkl}u_{k,l}u_{p,j}^* dR = \int_{C} C_{pjkl}u_k u_{p,j}^* n_l dC - \int_{R} C_{pjkl}u_k u_{p,jl}^* dR \tag{43}$$

Based on Eq. (43), Eq. (42) can be rewritten as

$$\int_{R} C_{pjkl}u_{k,lj}u_p^* \, dR - \int_{R} C_{pjkl}u_k u_{p,jl}^* dR = \int_{C} C_{pjkl}u_{k,l}u_p^* n_j dC - \int_{C} C_{pjkl}u_k u_{p,j}^* n_l dC \tag{44}$$

which can be written as

$$\int_{R} (L_{jl}u_k \cdot u_p^* - L_{jl}^* u_k^* \cdot u_p) dR = \int_{C} (G_{jl}u_k \cdot u_p^* - G_{jl}^* u_k^* \cdot u_p) dC \tag{45}$$

The boundary tractions are

$$t_p = C_{pjkl}u_{k,l}n_j = G_{jl}u_k \text{ and } t_p^* = C_{pjkl}u_{k,j}^* n_l = G_{jl}^* u_k^* \tag{46}$$

By using the symmetry relation of elasticity tensor, we obtain

$$L_{jl} = C_{pjkl}\aleph \frac{\partial^2}{\partial x_l \partial x_j} = C_{kjpl}\aleph \frac{\partial^2}{\partial x_j \partial x_l} = L_{jl}^* \tag{47}$$

$$G_{jl} = C_{pjkl}\aleph n_j \frac{\partial}{\partial x_l} = C_{kjpl}\aleph n_l \frac{\partial}{\partial x_j} = G_{jl}^* \tag{48}$$

Using Eqs. (46)–(48), the Eq. (45) can be reexpressed as

$$\int_{R} (C_{pjkl}u_{k,lj} \cdot u_p^* - C_{pjkl}u_{k,lj}^* \cdot u_p) dR = \int_{C} (t_p u_p^* - t_p^* u_p) dC \tag{49}$$

We define the fundamental solution $u_{mk}^*$ by the relation

$$L_{jl}u_{mk}^* = -\delta(x,\xi)\delta_{pm} \tag{50}$$

By modifying the weighting functions, Eq. (49) can be written as

$$\int_{R} (C_{pjkl}u_{k,lj} \cdot u_{mp}^* - C_{pjkl}u_{mk,lj}^* \cdot u_p) dR = \int_{C} (t_p u_{mp}^* - t_{mp}^* u_p) dC \tag{51}$$

From (39), (50) and (51), the representation formula may be written as

$$u_m(\xi) = \int_C \left(u^*_{mp}(x,\xi)t_p(x) - t^*_{mp}(x,\xi)u_p(x)\right)dC - \int_R u^*_{mp}(x,\xi)\rho b'_p(x)dR \tag{52}$$

Let

$$\rho b'_p = -\left(\left(D_{jl} + D_{pk} + \Lambda D_l\right)u_k + D_{pj}T_\alpha\right)$$

$$\approx \sum_{q=1}^{N} f^q_{pn} \propto^q_n = \sum_{q=1}^{N} \left(L_{jl}u^q_{kn}\right) \propto^q_n, D_{pk} = -\rho\delta_{pk}\frac{\partial^2}{\partial\tau^2} \tag{53}$$

The displacement particular solution may be defined as

$$u^q_{kn} = \delta_{kn}(r^2 + r^3) \tag{54}$$

Differentiation of (54) leads to.

$$u^q_{kn,l} = \delta_{kn}(2r + 3r^2)r_{,l}, \qquad u^q_{kn,lj} = \delta_{kn}((2 + 3r)\delta_{lj} + 3rr_{,j}r_{,l}) \tag{55}$$

Now, we obtain the traction particular solution $t^q_{pn}$ and source function $f^q_{pn}$ as

$$t^q_{pn} = C_{pjkl}u^q_{kn,l}n_j, \; L_{jl}u^q_{kn} = f^q_{pn} \tag{56}$$

The domain integral may be approximated as follows

$$\int_R u^*_{mp}\rho b'_p dR \approx \sum_{q=1}^{N}\left(\int_R \left(L_{jl}u^q_{kn}u^*_{mp}\right)dR\right)\propto^q_n \tag{57}$$

The use of (57) together with the dual reciprocity

$$\int_R \left(L_{jl}u^q_{kn}u^*_{mp} - L_{jl}u^*_{mk}u^q_{pn}\right)dR = \int_C \left(u^*_{mp}t^q_{pn} - t^*_{mp}u^q_{pn}\right)dC \tag{58}$$

Leads to

$$\int_R u^*_{mp}\rho b'_p dR = \sum_{q=1}^{N}\int_R \left(L_{jl}u^*_{mk}u^q_{pn}\right)dR + \int_C \left(u^*_{mp}t^q_{pn} - t^*_{mp}u^q_{pn}\right)dC\Big) \propto^q_n \tag{59}$$

From (50), we can write

$$\int_{R} L_{jl} u_{mk}^{*} u_{pn}^{q} \, dR = \int_{R} -\delta(x,\xi)\delta_{pm} u_{pn}^{q} \, dR = -u_{mn}^{q}(\xi) \tag{60}$$

By using (52), (59) and (60), we obtain

$$u_m(\xi) = \int_{C} \left(u_{mp}^{*} t_p - t_{mp}^{*} u_p\right) dC + \sum_{q=1}^{N} \left(u_{mn}^{q}(\xi) - \int_{C} \left(u_{mp}^{*} t_{pn}^{q} - t_{mp}^{*} u_{pn}^{q}\right) dC\right) \alpha_n^q \tag{61}$$

According to Fahmy [9–11], the right-hand side integrals of (61) can be reexpressed as

$$\int_{C} \left(u_{mp}^{*}(x,\xi) t_p(x) - t_{mp}^{*}(x,\xi) u_p(\xi)\right) dC = \lim_{\varepsilon \to 0} \int_{C-C_\varepsilon} \left(u_{mp}^{*}(x,\xi) t_p(x) - t_{mp}^{*}(x,\xi) u_p(\xi)\right) dC + \lim_{\varepsilon \to 0} \int_{C_\varepsilon} \left(u_{mp}^{*}(x,\xi) t_p(x) - t_{mp}^{*}(x,\xi) u_p(\xi)\right) dC_\varepsilon \tag{62}$$

and

$$\int_{C} \left(u_{mp}^{*}(x,\xi) t_{pn}^{q}(x) - t_{mp}^{*}(x,\xi) u_{pn}^{q}(\xi)\right) dC = \lim_{\varepsilon \to 0} \int_{C-C_\varepsilon} \left(u_{mp}^{*}(x,\xi) t_{pn}^{q}(x) - t_{mp}^{*}(x,\xi) u_{pn}^{q}(\xi)\right) dC + \lim_{\varepsilon \to 0} \int_{C_\varepsilon} \left(u_{mp}^{*}(x,\xi) t_{pn}^{q}(x) - t_{mp}^{*}(x,\xi) u_{pn}^{q}(\xi)\right) dC_\varepsilon \tag{63}$$

According to Fahmy [12], Guiggiani and Gigante [77] and Mantič [78] Eqs. (62) and (63) can respectively be expressed as

$$\int_{C} \left(u_{mp}^{*}(x,\xi) t_p(x) - t_{mp}^{*}(x,\xi) u_p(\xi)\right) dC = \int_{C} u_{mp}^{*} \, t_p d\Gamma - \oint_{C} u_p \, t_{mp}^{*} d\Gamma - c_{pj} u_p(\xi) \tag{64}$$

$$\int_{C} \left(u_{mp}^{*}(x,\xi) t_{pn}^{q}(x) - t_{mp}^{*}(x,\xi) u_{pn}^{q}(\xi)\right) dC = \int_{C} u_{mp}^{*} \, t_{pn}^{q} d\Gamma - \oint_{C} u_{pn}^{q} \, t_{mp}^{*} d\Gamma - c_{pj} u_{pn}^{q}(\varepsilon) \tag{65}$$

By using (64) and (65), the dual reciprocity boundary integral equation becomes

$$c_{pj}u_p(\varepsilon) + \oint_C u_p\, t^*_{mp} d\Gamma - \int_C u^*_{mp}\, t_p d\Gamma = \sum_{q=1}^{N}\left(c_{pj}u^q_{pn}(\varepsilon) + \oint_C u^q_{pn}\, t^*_{mp} d\Gamma - \int_C u^*_{mp}\, t^q_{pn} d\Gamma\right)\alpha^q_n \tag{66}$$

On the basis of isoparametric concept, we can write

$$\{u, t\} \approx \sum_{k=1}^{N} \varphi_k \{\breve{u}_k, \breve{t}_k\} = \Phi^T\{\breve{u}, \breve{t}\} \tag{67}$$

$$\{u^q, t^q\} \approx \sum_{k=1}^{N} \varphi_k \{\breve{u}^q_k, \breve{t}^q_k\} = \Phi^T\{\breve{u}^q, \breve{t}^q\} \tag{68}$$

By implementing the point collocation procedure and using (67) and (68), Eq. (66) may be reexpressed as

$$\zeta\breve{u}\text{-}\eta\breve{t} = \sum_{q=1}^{N}(\zeta\breve{u}^q\text{-}\eta\breve{t}^q)\alpha^q(\tau) \tag{69}$$

Let us suppose that

$$\breve{U} = [\breve{u}^1\ \breve{u}^2 \ldots \breve{u}^N] \tag{70}$$

$$\breve{\wp} = [\breve{t}^1\ \breve{t}^2 \ldots \breve{t}^N] \tag{71}$$

$$\alpha = [\alpha^1\ \alpha^2 \ldots \alpha^N]^T \tag{72}$$

We can write (69) as follows

$$\zeta\breve{u}(\tau)\text{-}\eta\breve{t}(\tau) = \left(\zeta\breve{U}\text{-}\eta\breve{\wp}\right)\alpha(\tau) \tag{73}$$

By using the point collocation procedure, $\boldsymbol{\alpha(\tau)}$ can be calculated from (53) as

$$\rho\breve{\ddot{u}}(\tau)\text{-}\rho\breve{b}(\tau) = F\alpha(\tau) \tag{74}$$

Now, from (74), we may derive

$$\alpha(\tau) = F^{-1}\left(\rho\breve{\ddot{u}}(\tau)\text{-}\rho\breve{b}(\tau)\right) \tag{75}$$

From (73) using (75) we have

$$\mathcal{M}\breve{\ddot{u}} + \zeta\breve{u} = \eta\breve{t}(\tau) + \breve{\mathcal{B}}(\tau) \tag{76}$$

where

$$\mho = \left(\eta\breve{\wp}\text{-}\zeta\breve{U}\right)F^{-1}, \qquad \mathcal{M} = \rho\mho, \qquad \breve{\mathcal{B}}(\tau) = \rho\mho\breve{b}(\tau). \tag{77}$$

By considering the following known $k$ and unknown $u$ superscripts nodal vectors

$$\{\breve{u}^k, \breve{t}^u\} \in C_3, \{\breve{u}^u, \breve{t}^k\} \in C_4 \tag{78}$$

Hence (76) may be written as

$$\begin{bmatrix} \mathcal{M}^{11} & \mathcal{M}^{12} \\ \mathcal{M}^{21} & \mathcal{M}^{22} \end{bmatrix} \begin{bmatrix} \ddot{\breve{u}}^k(\tau) \\ \ddot{\breve{u}}^u(\tau) \end{bmatrix} + \begin{bmatrix} \zeta^{11} & \zeta^{12} \\ \zeta^{21} & \zeta^{22} \end{bmatrix} \begin{bmatrix} \breve{u}^k(\tau) \\ \breve{u}^u(\tau) \end{bmatrix} = \begin{bmatrix} \eta^{11} & \eta^{12} \\ \eta^{21} & \eta^{22} \end{bmatrix} \begin{bmatrix} \breve{t}^k(\tau) \\ \breve{t}^u(\tau) \end{bmatrix} + \begin{bmatrix} \breve{\mathcal{B}}^1(\tau) \\ \breve{\mathcal{B}}^2(\tau) \end{bmatrix} \tag{79}$$

From the first row of (79), we can calculate the unknown fluxes $\breve{t}^u(\tau)$ as follows

$$\begin{aligned} &\breve{t}^u(\tau) \\ &= (\eta^{12})^{-1}\big[\mathcal{M}^{11}\ddot{\breve{u}}^k(\tau) + \mathcal{M}^{12}\ddot{\breve{u}}^u(\tau) + \zeta^{11}\breve{u}^k(\tau) \\ &+ \zeta^{12}\breve{u}^u(\tau)\text{-}\eta^{11}\breve{t}^k(\tau)\text{-}\breve{\mathcal{B}}^1(\tau)\big] \end{aligned} \tag{80}$$

From the second row of (79) and using (80) we get

$$\mathcal{M}^u\ddot{\breve{u}}^u(\tau) + \zeta^u\breve{u}^u(\tau) = Q^k(\tau) \tag{81}$$

where

$$\begin{aligned} Q^k(\tau) &= \breve{\mathcal{B}}^k(\tau) + \eta^k\breve{t}^k(\tau)\text{-}\mathcal{M}^k\ddot{\breve{u}}^k(\tau)\text{-}\zeta^k\breve{u}^k(\tau) \\ \mathcal{M}^u &= \mathcal{M}^{22}\text{-}\eta^{22}(\eta^{12})^{-1}\mathcal{M}^{12} \\ \zeta^u &= \zeta^{22}\text{-}\eta^{22}(\eta^{12})^{-1}\zeta^{12} \\ \mathcal{M}^k &= \mathcal{M}^{21}\text{-}\eta^{22}(\eta^{12})^{-1}\mathcal{M}^{11} \\ \zeta^k &= \zeta^{21}\text{-}\eta^{22}(\eta^{12})^{-1}\zeta^{11} \\ \breve{\mathcal{B}}^k(\tau) &= \mathcal{B}^2(\tau)\text{-}\eta^{22}(\eta^{12})^{-1}\mathcal{B}^1(\tau) \end{aligned} \tag{82}$$

Eq. (81) can be written at $(n+1)$ time step as

$$\mathcal{M}^u\ddot{\breve{u}}^u_{n+1}(\tau) + \zeta^u\breve{u}^u_{n+1}(\tau) = Q^k_{n+1}(\tau) \tag{83}$$

where

$$Q^k_{n+1}(\tau) = \breve{\mathcal{B}}^k_{n+1}(\tau) + \eta^k\breve{t}^k_{n+1}(\tau)\text{-}\mathcal{M}^k\ddot{\breve{u}}^k_{n+1}(\tau)\text{-}\zeta^k\breve{u}^k_{n+1}(\tau) \tag{84}$$

In order to solve (83), The implicit backward finite difference scheme has been applied based on the Houbolt's algorithm and the following approximations

$$\dot{\breve{u}}_{n+1} \approx \frac{1}{6\Delta t}(11\breve{u}_{n+1}\text{-}18\breve{u}_n + 9\breve{u}_{n\text{-}1}\text{-}2\breve{u}_{n\text{-}2}) \tag{85}$$

$$\ddot{\breve{u}}_{n+1} \approx \frac{1}{\Delta t^2}(2\breve{u}_{n+1}\text{-}5\breve{u}_n + 4\breve{u}_{n\text{-}1}\text{-}\breve{u}_{n\text{-}2}) \tag{86}$$

By using (85) and (86), we have from (83)

$$\varsigma^u \breve{u}^u_{n+1}(\tau) = \mathbb{Q}^k_{n+1}(\tau) \tag{87}$$

In which.

$$\varsigma^u = \frac{2\mathcal{M}^u}{\Delta t^2} + \zeta^u \tag{88}$$

$$\mathbb{Q}^k_{n+1} = Q^k_{n+1} + \frac{\mathcal{M}^u}{\Delta t^2}(5\breve{u}_n\text{-}4\breve{u}_{n\text{-}1} + \breve{u}_{n\text{-}2}) \tag{89}$$

We implement the successive over-relaxation (SOR) of Golub and Van Loan [79] for solving (87) to obtain $\breve{u}^u_{n+1}$. Then, the unknown $\dot{\breve{u}}^u_{n+1}$ and $\ddot{\breve{u}}^u_{n+1}$ can be obtained from (76) and (77), respectively. By using the procedure of Bathe [80], we obtain the traction vector $t^u_{n+1}$ from (73).

## 5. Numerical results and discussion

The BEM that has been used in the current paper can be applicable to a wide variety of FGA structures problems associated with the proposed theory of three temperatures nonlinear uncoupled magneto-thermoelasticity. In order to evaluate the influence of graded parameter on the three temperatures and displacements, the numerical results are carried out and depicted graphically for homogeneous ($m = 0$) and functionally graded ($m = 0.5$ and $1.0$) structures.

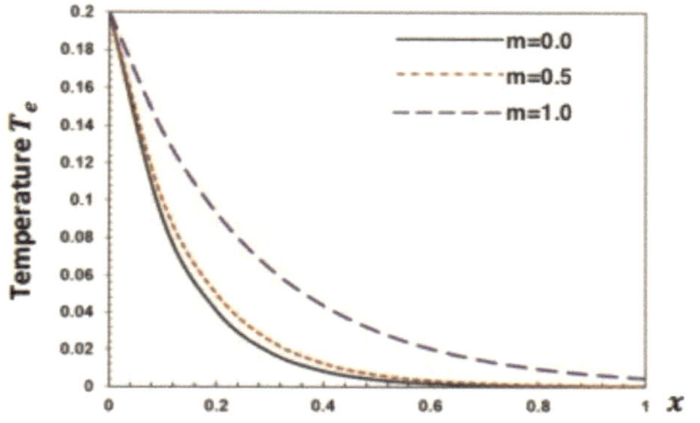


**Figure 2.**
*Variation of the electron temperature $T_e$ through the thickness coordinate x.*

**Figures 2–4** show the distributions of the three temperatures $T_e$, $T_i$ and $T_p$ through the thickness coordinate $Ox$. It was shown from these figures that the three temperatures increase with increasing value of graded parameter m.

**Figures 5** and **6** show the distributions of the displacements $u_1$ and $u_2$ through the thickness coordinate Ox. It was noticed from these figures that the displacement components increase with increasing value of graded parameter m.

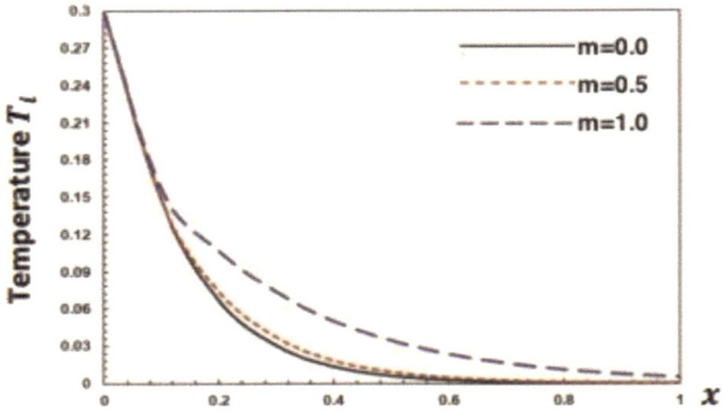


**Figure 3.**
*Variation of the ion temperature $T_i$ through the thickness coordinate x.*

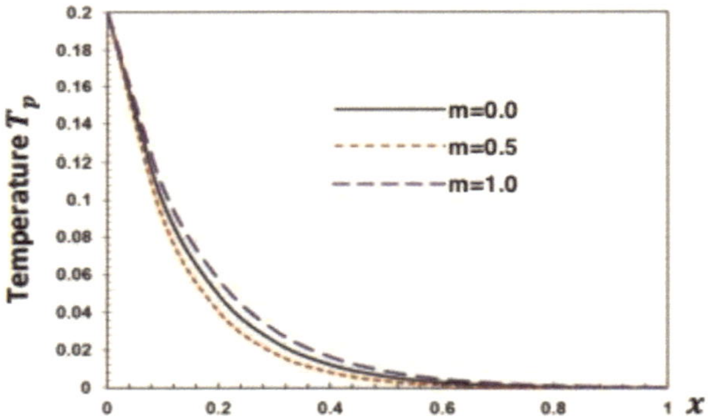


**Figure 4.**
*Variation of the photon temperature $T_p$ through the thickness coordinate x.*

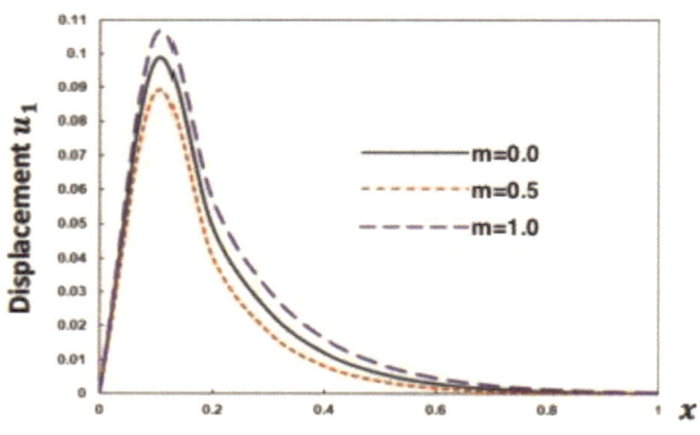


**Figure 5.**
*Variation of the displacement $u_1$ through the thickness coordinate x.*

**Figures 7** and **8** show the distributions of the displacements $u_1$ and $u_2$ with the time for boundary element method (BEM), finite difference method (FDM) and finite element method (FEM) to demonstrate the validity and accuracy of our proposed technique. It is noted from numerical results that the BEM obtained results are agree quite well with those obtained using the FDM of Pazera and Jędrysiak [81] and FEM of Xiong and Tian [82] results based on replacing heat conduction with three-temperature heat conduction.

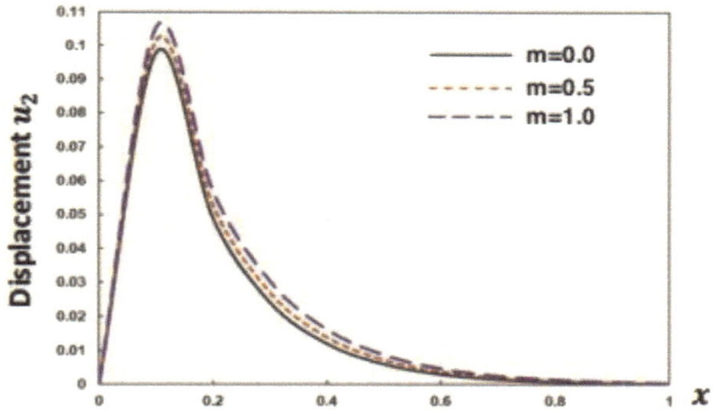


**Figure 6.**
*Variation of the displacement $u_2$ through the thickness coordinate x.*

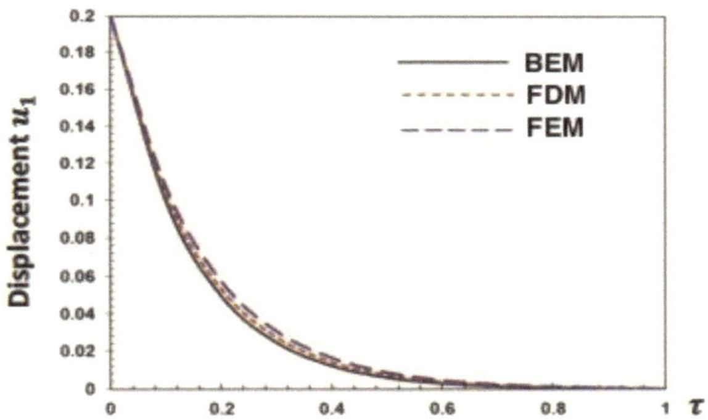


**Figure 7.**
*Variation of the displacement $u_1$ with time $\tau$.*

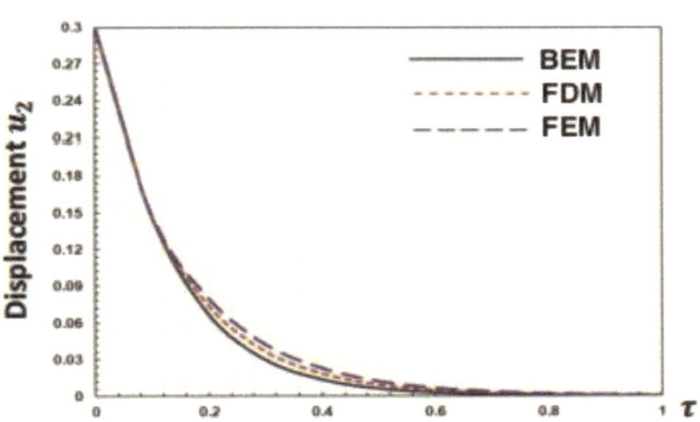


**Figure 8.**
*Variation of the displacement $u_2$ with time $\tau$.*

## 6. Conclusion

The main aim of this article is to introduce a new fractional-order theory called nonlinear uncoupled magneto-thermoelasticity theory involving three temperatures for FGA structures and new boundary element technique for solving problems related to the proposed theory. Since the nonlinear three temperatures radiative heat conduction equation is independent of the displacement field, we first determine the temperature field using the BEM, then based on the known temperature field, the displacement field is obtained by solving the move equation using the BEM. It can be seen from the numerical results that the graded parameter had a significant effect on the temperatures and displacements through the thickness of the functionally graded structures. Since there are no available results for the considered problem. So, some literatures may be considered as special cases from the considered problem based on replacing the heat conduction by three temperatures radiative heat conduction. The numerical results demonstrate the validity and accuracy of our proposed model. From the proposed BEM technique that has been performed in the present paper, it is possible to conclude that the proposed BEM should be applicable to any FGM uncoupled magneto-thermoelastic problem with three-temperature. BEM is more efficient, accurate and easy to use than FDM or FEM, because it only needs to solve the unknowns on the boundaries and BEM users need only to deal with real geometry boundaries. Also, BEM is reducing the computational cost of its solver. The present numerical results for our complex problem may provide interesting information for computer scientists, designers of new FGM materials and researchers in FGM science as well as for those working on the development of new functionally graded structures.

## Author details

Mohamed Abdelsabour Fahmy
Faculty of Computers and Informatics, Suez Canal University, Ismailia, Egypt

*Address all correspondence to: mohamed_fahmy@ci.suez.edu.eg

## References

[1] Reddy JN. Analysis of functionally graded plated. International Journal for Numerical Methods in Engineering. 2000;**47**:663-384

[2] Reddy JN. Mechanics of Laminated Composite Plates and Shells. 2nd ed. Boca Raton, FL: CRC Press, LLC; 2004

[3] Abd-Alla AM, Fahmy MA, El-Shahat TM. Magneto-thermo-elastic problem of a rotating non-homogeneous anisotropic solid cylinder. Archive of Applied Mechanics. 2008;**78**:135-148

[4] Fahmy MA. Thermoelastic stresses in a rotating non- homogeneous anisotropic body. Numerical Heat Transfer, Part A: Applications. 2008;**53**: 1001-1011

[5] Fahmy MA, El-Shahat TM. The effect of initial stress and inhomogeneity on the thermoelastic stresses in a rotating anisotropic solid. Archive of Applied Mechanics. 2008;**78**:431-442

[6] Fahmy MA. A time-stepping DRBEM for magneto-thermo-viscoelastic interactions in a rotating nonhomogeneous anisotropic solid. International Journal of Applied Mechanics. 2011;**3**:1-24

[7] El-Naggar AM, Abd-Alla AM, Fahmy MA, Ahmed SM. Thermal stresses in a rotating non-homogeneous orthotropic hollow cylinder. Heat and Mass Transfer. 2002;**39**:41-46

[8] Abd-Alla AM, El-Naggar AM, Fahmy MA. Magneto-thermoelastic problem in non-homogeneous isotropic cylinder. Heat and Mass Transfer. 2003;**39**: 625-629

[9] Fahmy MA. A time-stepping DRBEM for the transient magneto-thermo-visco-elastic stresses in a rotating non-homogeneous anisotropic solid. Engineering Analysis with Boundary Elements. 2012;**36**:335-345

[10] Fahmy MA. Transient magneto-thermoviscoelastic plane waves in a non-homogeneous anisotropic thick strip subjected to a moving heat source. Applied Mathematical Modelling. 2012;**36**:4565-4578

[11] Fahmy MA. The effect of rotation and inhomogeneity on the transient magneto-thermoviscoelastic stresses in an anisotropic solid. ASME Journal of Applied Mechanics. 2012;**79**:1015

[12] Fahmy MA. Transient magneto-thermo-elastic stresses in an anisotropic viscoelastic solid with and without a moving heat source. Numerical Heat Transfer, Part A: Applications. 2012;**61**: 633-650

[13] Jabbari M, Sohrabpour S, Elsami MR. Mechanical and thermal stresses in a functionally graded hollow cylinder due to radially symmetric loads. International Journal of Pressure Vessels and Piping. 2002;**79**:493-497

[14] Bhandari M, Ashirvad KP. Comparison of functionally graded material plate with metal and ceramic plate under transverse load for various boundary conditions. In: International Journal of Computer Applications, National Conference in Advances in Technology & Applied Sciences (NCATAS-2014); September 2014. pp. 25-29

[15] Sburlati R, Bardella L. Three-dimensional elastic solutions for functionally graded circular plates. European Journal of Mechanics - A/Solids. 2011;**30**:219-235

[16] Li XY, Li PD, Kang GZ, Pan DZ. Axisymmetric thermo-elasticity field in a functionally graded circular plate of

transversely isotropic material. Mathematics and Mechanics of Solids. 2012;**18**:464-475

[17] Fahmy MA. A 2-D DRBEM for generalized magneto-thermo-viscoelastic transient response of rotating functionally graded anisotropic thick strip. International Journal of Engineering and Technology Innovation. 2013;**3**:70-85

[18] Fahmy MA. The DRBEM solution of the generalized magneto-thermo-viscoelastic problems in 3D anisotropic functionally graded solids. In: Proceedings of the 5th International Conference on Coupled Problems in Science and Engineering (Coupled Problems 2013); 17–19 June 2013; Ibiza, Spain. pp. 862-872

[19] Fahmy MA, Salem AM, Metwally MS, Rashid MM. Computer implementation of the DRBEM for studying the generalized Thermoelastic responses of functionally graded anisotropic rotating plates with one relaxation time. International Journal of Applied Science and Technology. 2013; **3**:130-140

[20] Fahmy MA, Salem AM, Metwally MS, Rashid MM. Computer implementation of the DRBEM for studying the classical uncoupled theory of thermoelasticity of functionally graded anisotropic rotating plates. International Journal of Engineering Research and Applications. 2013;**3**: 1146-1154

[21] Fahmy MA, Salem AM, Metwally MS, Rashid MM. Computer implementation of the Drbem for studying the classical coupled thermoelastic responses of functionally graded anisotropic plates. Physical Science International Journal. 2014;**4**: 674-685

[22] Fahmy MA. Numerical modeling of coupled Thermoelasticity with relaxation times in rotating FGAPs subjected to a moving heat source. Physical Science International Journal. 2016;**9**:1-13

[23] Fahmy MA. DRBEM sensitivity analysis and shape optimization of rotating magneto-thermo-viscoelastic FGA structures using Golden-section search algorithm based on uniform Bicubic B-splines. Journal of Advances in Mathematics and Computer Science. 2017;**25**:1-20

[24] Fahmy MA. A predictor-corrector time-stepping Drbem for shape design sensitivity and optimization of multilayer FGA structures. Transylvanian Review. 2017;**XXV**: 5369-5382

[25] Mojdehi AR, Darvizeh A, Basti A, Rajabi H. Three dimensional static and dynamic analysis of thick functionally graded plates by the meshless local Petrov-Galerkin (MLPG) method. Engineering Analysis with Boundary Elements. 2011;**35**: 1168-1180

[26] Rekik M, El-Borgi S, Ounaies Z. An embedded mixed-mode crack in a functionally graded magnetoelectroelastic infinite medium. International Journal of Solids and Structures. 2012;**49**:835-845

[27] Reddy J. Analysis of functionally graded plates. International Journal for Numerical Methods in Engineering. 2000;**47**:663-684

[28] Zhong Z, Shang E. Three-dimensional exact analysis of a simply supported functionally gradient piezoelectric plate. International Journal of Solids and Structures. 2003;**40**: 5335-5352

[29] Ramirez F, Heyliger PR, Pan E. Static analysis of functionally graded elastic anisotropic plates using a discrete

layer approach. Composites Part B: Engineering. 2006;**37**:10-20

[30] Li X, Ding H, Chen W. Axisymmetric elasticity solutions for a uniformly loaded annular plate of transversely isotropic functionally graded materials. Acta Mechanica. 2008;**196**:139-159

[31] Kashtalyan M, Menshykova M. Three-dimensional elasticity solution for sandwich panels with a functionally graded core. Composite Structures. 2009;**87**:36-43

[32] Yang B, Ding H, Chen W. Elasticity solutions for a uniformly loaded rectangular plate of functionally graded materials with two opposite edges simply supported. Acta Mechanica. 2009;**207**:245-258

[33] Vel SS, Batra RC. Three-dimensional analysis of transient thermal stresses in functionally graded plates. International Journal of Solids and Structures. 2003;**40**:7181-7196

[34] Zhang XZ, Kitipornchai S, Liew KM, Lim CW, Peng LX. Thermal stresses around a circular hole in a functionally graded plate. Journal of Thermal Stresses. 2003;**26**:379-390

[35] Ohmichi M, Noda N. Plane thermoelastic problem in a functionally graded plate with an oblique boundary to the functional graded direction. Journal of Thermal Stresses. 2007;**30**:779-799

[36] Ootao Y, Tanigawa Y. Transient thermoelastic analysis for a laminated composite strip with an interlayer of functionally graded material. Journal of Thermal Stresses. 2009;**32**:1181-1197

[37] Obata Y, Noda N. Transient thermal stresses in a plate of functionally gradient material. In: Holt JB et al., editors. Ceramic Trans. Functionally Gradient Materials. Vol. 34. Westerville, Ohio, USA: The American Ceramic Society; 1993. pp. 403-410

[38] Tanigawa Y, Morishita H, Ogaki S. Derivation of systems of fundamental equations for a three-dimensional thermoelastic field with nonhomogeneous material properties and its application to a semi-infinite body. Journal of Thermal Stresses. 1999;**22**:689-711

[39] Cheng ZQ, Batra R. Three-dimensional thermoelastic deformations of a functionally graded elliptic plate. Composites Part B: Engineering. 2000;**31**:97-106

[40] Reddy J, Cheng ZQ. Three-dimensional thermomechanical deformations of functionally graded rectangular plates. European Journal of Mechanics—A/Solids. 2001;**20**:841-855

[41] Sankar BV, Tzeng JT. Thermal stresses in functionally graded beams. AIAA Journal. 2002;**40**:1228-1243

[42] Ootao Y, Tanigawa Y. Three-dimensional solution for transient thermal stresses of an orthotropic functionally graded rectangular plate. Composite Structures. 2007;**80**:10-20

[43] Jabbari M, Dehbani H, Eslami M. An exact solution for classic coupled thermoelasticity in spherical coordinates. Journal of Pressure Vessel Technology. 2010;**132**:031201

[44] Jabbari M, Dehbani H, Eslami M. An exact solution for classic coupled thermoelasticity in cylindrical coordinates. Journal of Pressure Vessel Technology. 2011;**133**:051204

[45] Akbarzadeh A, Abbasi M, Eslami M. Coupled thermoelasticity of functionally graded plates based on the third-order shear deformation theory. Thin-Walled Structures. 2012;**53**:141-155

[46] Fahmy MA. A three-dimensional generalized magneto-thermo-viscoelastic problem of a rotating functionally graded anisotropic solids with and without energy dissipation. Numerical Heat Transfer, Part A: Applications. 2013;**63**:713-733

[47] Fahmy MA. A 2D time domain DRBEM computer model for magneto-thermoelastic coupled wave propagation problems. International Journal of Engineering and Technology Innovation. 2014;**4**:138-151

[48] Fahmy MA, Salem AM, Metwally MS, Rashid MM. Computer implementation of the DRBEM for studying the generalized thermo elastic responses of functionally graded anisotropic rotating plates with two relaxation times. British Journal of Mathematics & Computer Science. 2014;**4**:1010-1026

[49] Fahmy MA, Al-Harbi SM, Al-Harbi BH. Implicit time-stepping DRBEM for design sensitivity analysis of magneto-thermo- elastic FGA structure under initial stress. American Journal of Mathematical and Computational Sciences. 2017;**2**:55-62

[50] Mazarei Z, Nejad MZ, Hadi A. Thermo-Elasto-plastic analysis of thick-walled spherical pressure vessels made of functionally graded materials. International Journal of Applied Mechanics. 2016;**8**:1650054

[51] Su Z, Jin G, Wang L, Wang D. Thermo-mechanical vibration analysis of size-dependent functionally graded micro-beams with general boundary conditions. International Journal of Applied Mechanics. 2018;**10**:1850088

[52] Tomar SS, Talha M. Thermo-mechanical buckling analysis of functionally graded skew laminated plates with initial geometric imperfections. International Journal of Applied Mechanics. 2018;**10**:1850014

[53] Ueda S, Okada M, Nakaue Y. Transient thermal response of a functionally graded piezoelectric laminate with a crack normal to the bimaterial interface. Journal of Functionally Graded Materials. 2017;**31**: 14-23

[54] Valizadeh N, Natarajan S, Gonzalez-Estrada OA, Rabczuk T, Bui T, Bordas SPA. NURBS-based finite element analysis of functionally graded plates: Static bending, vibration, buckling and flutter. Composite Structures. 2013;**99**: 309-326

[55] Bhardwaj G, Singh IV, Mishra BK, Bui TQ. Numerical simulation of functionally graded cracked plates using NURBS based XIGA under different loads and boundary conditions. Composite Structures. 2015;**126**:347-359

[56] Sladek J, Sladek V, Zhang C. A local BIEM for analysis of transient heat conduction with nonlinear source terms in FGMs. Engineering Analysis with Boundary Elements. 2004;**28**:1-11

[57] Sladek J, Sladek V, Zhang C. Transient heat conduction analysis in functionally graded materials by the meshless local boundary integral equation method. Computational Materials Science. 2003;**28**:494-504

[58] Sladek J, Sladek V, Krivacek J, Zhang C. Local BIEM for transient heat conduction analysis in 3-D axisymmetric functionally graded solids. Computational Mechanics. 2003; **32**:169-176

[59] Gao XW, Zhang C, Sladek J, Sladek V. Fracture analysis of functionally graded materials by a BEM. Composites Science and Technology. 2008;**68**: 1209-1215

[60] Fahmy MA. Implicit-explicit time integration DRBEM for generalized magneto-thermoelasticity problems of rotating anisotropic viscoelastic

functionally graded solids. Engineering Analysis with Boundary Elements. 2013; **37**:107-115

[61] Fahmy MA. Generalized magneto-thermo-viscoelastic problems of rotating functionally graded anisotropic plates by the dual reciprocity boundary element method. Journal of Thermal Stresses. 2013;**36**: 1-20

[62] Fahmy MA. A computerized DRBEM model for generalized magneto-thermo-visco-elastic stress waves in functionally graded anisotropic thin film/substrate structures. Latin American Journal of Solids and Structures. 2014;**11**: 386-409

[63] Fahmy MA. Boundary element solution of 2D coupled problem in anisotropic piezoelectric FGM plates. In: Proceedings of the 6th International Conference on Computational Methods for Coupled Problems in Science and Engineering (Coupled Problems 2015); 18–20 May 2015; Venice, Italy; pp. 382-391

[64] Fahmy MA. 3D DRBEM modeling for rotating initially stressed anisotropic functionally graded piezoelectric plates. In: Proceedings of the 7th European Congress on Computational Methods in Applied Sciences and Engineering (ECCOMAS 2016); 5–10 June 2016; Crete Island, Greece. pp. 7640-7658

[65] Fahmy MA. The effect of anisotropy on the structure optimization using Golden-section search algorithm based on BEM. Journal of Advances in Mathematics and Computer Science. 2017;**25**:1-18

[66] Fahmy MA. A time-stepping DRBEM for 3D anisotropic functionally graded piezoelectric structures under the influence of gravitational waves. In: Proceedings of the 1st GeoMEast International Congress and Exhibition (GeoMEast 2017); 15–19 July 2017; Sharm El Sheikh, Egypt. Facing the Challenges in Structural Engineering, Sustainable Civil Infrastructures. 2017. pp. 350-365

[67] Fahmy MA. Shape design sensitivity and optimization for two-temperature generalized magneto-thermoelastic problems using time-domain DRBEM. Journal of Thermal Stresses. 2018;**41**: 119-138

[68] Fahmy MA. Shape design sensitivity and optimization of anisotropic functionally graded smart structures using bicubic B-splines DRBEM. Engineering Analysis with Boundary Elements. 2018;**87**:27-35

[69] Fahmy MA. Modeling and optimization of anisotropic viscoelastic porous structures using CQBEM and moving asymptotes algorithm. Arabian Journal for Science and Engineering. 2019;**44**:1671-1684

[70] Fahmy MA. Boundary element algorithm for modeling and simulation of dual phase lag bioheat transfer and biomechanics of anisotropic soft tissues. International Journal of Applied Mechanics. 2018;**10**:1850108

[71] Fahmy MA. A new computerized boundary element algorithm for cancer modeling of cardiac anisotropy on the ECG simulation. Asian Journal of Research in Computer Science. 2018;**2**: 1-10

[72] Fahmy MA. Boundary element modeling and simulation of Biothermomechanical behavior in anisotropic laser-induced tissue hyperthermia. Engineering Analysis with Boundary Elements. 2019;**101**: 156-164

[73] Fahmy MA. Computerized Boundary Element Solutions for Thermoelastic Problems: Applications to Functionally Graded Anisotropic

Structures. Saarbrücken: LAP Lambert Academic Publishing; 2017

[74] Fahmy MA. Boundary Element Computation of Shape Sensitivity and Optimization: Applications to Functionally Graded Anisotropic Structures. Saarbrücken: LAP Lambert Academic Publishing; 2017

[75] Cattaneo C. Sur une forme de i'equation de la chaleur elinant le paradox d'une propagation instantance. Comptes Rendus de l'Académie des Sciences. 1958;**247**:431-433

[76] Wrobel LC. The Boundary Element Method, Applications in Thermos-Fluids and Acoustics. New York: Wiley; 2002

[77] Guiggiani M, Gigante A. A general algorithm for multidimensional Cauchy principal value integrals in the boundary element method. ASME Journal of Applied Mechanics. 1990;**57**:906-915

[78] Mantič V. A new formula for the C-matrix in the Somigliana identity. Journal of Elasticity. 1993;**33**:191-201

[79] Golub GH, Van Loan CF. Matrix Computations. Oxford: North Oxford Academic; 1983

[80] Bathe KJ. Finite Element Procedures. Englewood Cliffs: Prentice-Hall; 1996

[81] Pazera E, Jędrysiak J. Effect of microstructure in thermoelasticity problems of functionally graded laminates. Composite Structures. 2018; **202**:296-303

[82] Xiong QL, Tian XG. Generalized magneto-thermo-microstretch response during thermal shock. Latin American Journal of Solids and Structures. 2015; **12**:2562-2580

Chapter 3

# DC Conductivity of Activated Carbon Filled Epoxy Gradient Composites

*Archana Nigrawal*

## Abstract

This chapter reports the DC conductivity behavior of activated carbon powder filled epoxy gradient composites. Gradient composites are the composite materials in which the there is gradually variation in some direction to achieve gradient in properties. Graded materials are generally defined as the materials, which exhibit variable functional performance with location and show continuous variations in morphology and composition. Functionally graded metal matrix composites have been of great practical importance. Graded metal matrix composite have gradual compositional variations from ceramic at one surface to metal at the other, leading to special advantages of smooth transition in thermal stresses across the thickness and minimized stress concentration at the interface of two dissimilar materials. Therefore graded metal matrix composites are finding applications in aggressive environments with steep temperature gradients such as turbine components and rocket nozzles. Since the properties of material in FGMs are variable across the entire material, and depends on the spatial position of the material. Functionally graded materials are designed with varying properties such as changing their chemical properties, changing mechanical, magnetic, thermal, and electrical properties. Now a days there are FGMs designed as stepwise-graded materials, while others are fabricated to have continuous-graded materials depending on their areas of application.

**Keywords:** activated carbon powder, epoxy, gradient, DC conductivity, composites

## 1. Introduction

Activated carbon powder filled epoxy composites having 3 wt.% of activated carbon powder and epoxy resin have been developed. DC conductivity measurements are conducted on the graded composites by using an Electro-meter in the temperature range from 28 to 150°C. DC conductivity increases with the increase of distance in the direction of centrifugal force, which shows the formation of graded structure with the composites. DC conductivity increases on increase in activated carbon powder content. Activation energy was calculated and showed ionic conduction in the composites. Polymers with conductive fillers have many applications in solid state devices, mostly in fabrication of polymer light-emitting diodes, microelectronic components, optical displays as battery and fuel cell electrodes, antistatic media, corrosion-resistant materials, etc. [1–3]. The advantage of using conductive filler is that it is possible to control their electrical conductivity.

Besides good electrical conductivity and optical characteristics, conducting polymers have several other advantages as plasticity, low cost, lightweight and ease of fabrication [4–6].

Functionally graded composites are a novel class of composites which shows unique properties. Their graded property can be used as medical implants, for thermal protection in space vehicles, and they can be used as thermoelectric converter for energy conservation. Due to their versatile nature, they are widely used in nano, opto-electronic and thermoelectric materials also. Future applications of carbon nanotubes (CNT) reinforced functionally graded composite materials (FGCM) is expected to unique material having a wide range of possibilities in various areas such as aerospace, energy, automobile, medicine and structural industry. These materials can be used as gas adsorbents, probes, chemical sensors, nanopipes, nano-reactors, etc.

FGM can be used according the desired applications in biomedical application, as implants in human body to function properly without destroying the surrounding tissues.

Nigrawal and Chand has studied the dielectric properties of activated carbon filled epoxy composites and reported that small values of activation energy obtained at higher frequencies suggested that the conduction in the composite was due to hopping of charge carriers [7].

Epoxy resins are used as a thermosetting polymer matrix for the preparation of the conductive polymer composites [7–9]. Most of the electrically conductive polymer composites consist of carbon fibres or carbon black. A degree of conductivity is achieved when the concentration of the fillers is high enough so as to form a conductive network within the polymer matrix and such critical concentration, is known as percolation threshold. Conductivity of polymers containing conductive fillers, depends on the size and shape of the filler particles, spatial distribution and the contact resistance [10–14] and also determines the conditions of charge transport. Polymers have wide applications as electrical and electronic materials in the field of electro-photography and opto-electronics [15, 16] and can also be used as interfacial barrier layers and protective coatings. However, prior to using these materials in these specialized applications it is essential to know fundamental properties such as the mechanisms of electrical conduction, charge storage, etc.

It is well known that the addition of nano size additives into polymers has paved way for advanced technologies, such as in electrochemical displays, sensors, catalysis and redox capacitors, etc. [7]. For instance graphene based polymer composites possess potential applications in radiation and electromagnetic shielding, antistatic, shrinkage and corrosion-resistant coatings, and other mechanical and functional attributes such as stiffness, barrier, conducting capabilities, light emitting devices, batteries and other functional applications. Other potential applications could be for high temperature conducting adhesive, or for the bipolar plates for polymer electrolyte membrane fuel cells. Several carbon additives have also been utilized to enhance the properties of a polymer, the most popular being carbon blacks, carbon nanotubes [15–19].

During curing of the thermoset matrix, an internal stress comes and increases the pressure between and decreases the contact resistance. Development and modification of carbonaceous materials in necessary to increase the specific energy and power of supercapacitors, by controlling the pore size distribution, introducing electroactive metallic particles or electroconducting polymers [17–19].

It was reported that conductive polymer composite sensor array made from carbon black with polymers [20, 21].

An activated carbon has extended surface area, microporous structure, high adsorption capacity and high degree of surface reactivity. Activated carbon can be used into various structures such as fill-form, felt-form and fabric-form for various applications [22].

Different types of activated carbon powder and as well as activated carbon fibers with surface area in the range of 86–3000 $m^2/g$ were used for electrical double-layer capacitors [23, 24].

The conductivity of conductive filler polymer systems depends on the filler type, concentration, structure, surface properties, and conductivity properties of the matrix, distribution of particles in the matrix, contacts between particles, and particle orientation. It is well known that carbon black (CB) particles with a larger structure may render a relative high conductivity [25].

In a recent study effects of carbon nanotube on volume fractions, slenderness ratio, and core-to-face sheet thickness ratio on free vibration behavior of sandwich beams with functionally graded carbon nanotube-reinforced composite was reported. Numerical results were also reported to compare the behavior of sandwich beams of carbon nanotube-reinforced composite face sheets to those with functionally graded carbon nanotube-reinforced composite face sheets [26].

Flexural properties of epoxy nano composites increased on addition of 1 and 5% vol nano-activated carbon as compared to neat epoxy. A noticeable increment in flexural strength and modulus, was observed on addition of 1 and 5% vol nano-activated carbon as compared to pure epoxy. The effect of potassium hydroxide and phosphoric acid treatment on activated carbon epoxy nano composites were also investigated. The flexural toughness of both the composites were in range between 0.79 and 0.92 J. It was reported that the flexural strength of activated carbon-bamboo stem, activated carbon-oil palm, and activated carbon-coconut shell reinforced epoxy nanocomposites showed almost same value in case of 5% potassium hydroxide activated carbon [27].

In case of many FGMs components, properties varies in thickness direction [28]. However, in many modern applications these material have variable properties in both thickness and axial directions [28]. In a paper a gradient material in which properties varies in both aspects are developed and studied [28, 29]. Such smart materials are known as bidirectional functionally gradient (BDFGMs) materials. In which laser metal deposition based AM technique was used [29].

## 2. Materials and methods

Activated carbon powder used in this study was obtained from Ranbaxy Fine Chemicals, New Delhi, India. Epoxy used in this study was obtained from Hindustan Advanced Materials (India) Pvt. Ltd., Chakala (east), Mumbai, India. Activated carbon powder filled epoxy gradient composites have been developed by using centrifugation process. In this process centrifugal force is applied in the X direction. Gradient samples are prepared from the activated carbon powder filled mix having 3 wt.% of activated carbon powder. Activated carbon powder was added to a mix of epoxy resin and hardener (10:8). Total mix was thoroughly stirred with the help of a glass rod. Details of set up and process of making gradient composites are as reported in earlier patent (Chand and Hashmi) [27]. Total mix was thoroughly stirred with the help of a glass rod at 24°C for 2 min. The total mix was kept in a cylindrical mould to make graded sample. The sample with mould was rotated at 800 ± 50 RPM at a radius of 130 mm. Graded sample pin was removed from the mould after post curing at room temperature for 24 h. Composite pin was sliced into four pieces starting from centre to periphery and designated as sample 1, 2, 3 and 4, respectively. Samples were coated on both the sides by air drying type silver paint before the electrical measurements.

Density of activated carbon powder filled epoxy gradient samples was measured by using a Mettler Toledo precision balance.

### 2.1 Resistivity measurements

Resistance (R) values of activated carbon powder filled epoxy gradient samples were measured by using a kiethley electrometer model 610°C in the temperature range ranging from 28 to 150°C. Heating rate was kept constant at 1°C/min.

DC conductivity ($\sigma_{DC}$) values were calculated by using the following relation

$$\rho = R^{*} A/l \tag{1}$$

where R is the resistance value of the sample; A ($cm^2$) is the area of the electrodes; and l (cm) is the thickness of the sample.

Conductivity was calculated by using the following formula.

$$\sigma DC = 1/\rho \tag{2}$$

### 2.2 Density measurements

Densities of activated carbon filled epoxy resin composites were determined by using a Mettler Toledo precision balance.

## 3. Results and discussion

**Figure 1** shows the schematic view of the gradient composite sample prepared using a mix of activated carbon powder and epoxy. This schematic diagram shows the distribution of activated carbon powder in the composite.

**Table 1** lists the density of activated carbon powder filled epoxy gradient composites at different distances from periphery. This shows the increase of distance from periphery decreases the density of the composite. This is due to the decrease in activated carbon content. **Figure 2** shows the variation of DC conductivity with temperature for activated carbon powder filled epoxy sample 1. In this plot DC conductivity increased at 108°C then at 112°C it became constant up to 128°C and then there is a sudden increase in DC conductivity after 128°C, and it increases up to 150°C.

**Figure 3** shows the variation of DC conductivity with temperature for activated carbon powder filled epoxy gradient composites sample 2. This plot shows that up to 98°C there is no increase in DC conductivity. After 106°C there is a sudden increase in DC conductivity with increase of temperature and a peak was found at 138°C temperature. This plot shows that there is an increase in DC conductivity with the increase in activated carbon powder content at all temperatures. Another important

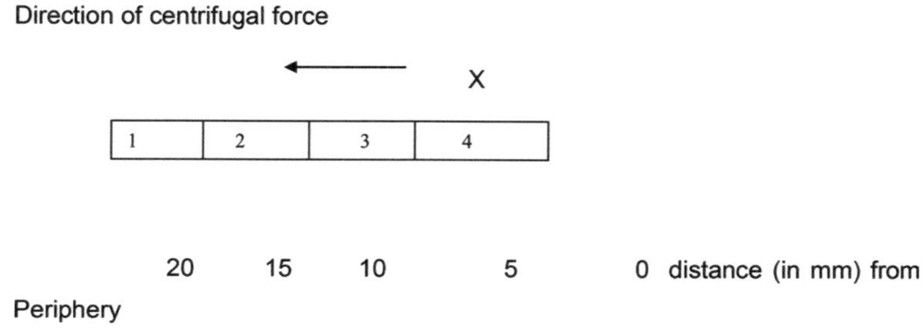


**Figure 1.**
*Schematic diagram of activated carbon distribution in four sections.*

| Sample no. | Density (g/cc) |
|---|---|
| Sample 1 | 1.000 |
| Sample 2 | 1.05317 |
| Sample 3 | 1.06169 |
| Sample 4 | 1.0999 |

**Table 1.**
*Lists the density (ρ) values of gradient composite at different distances.*

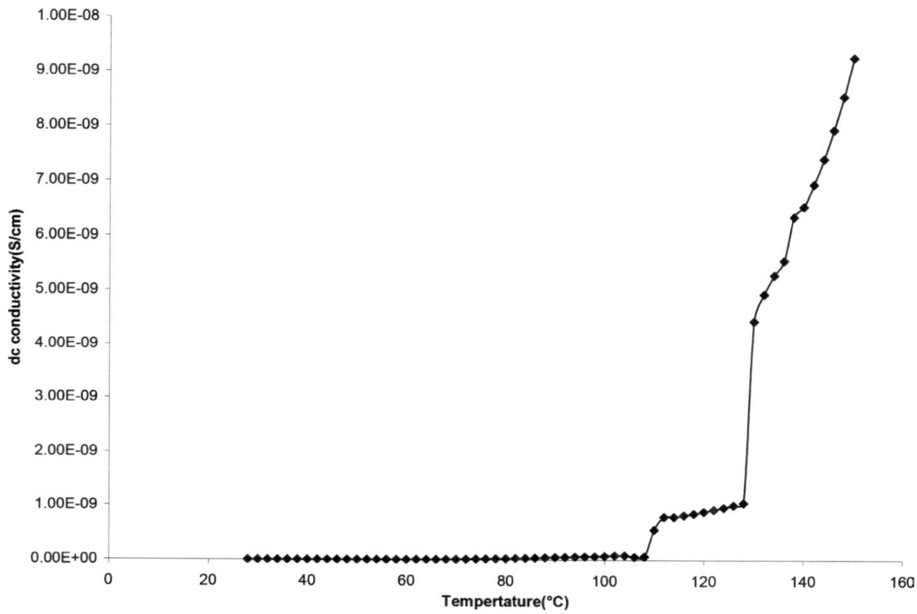


**Figure 2.**
*Variation of DC conductivity with temperature for activated carbon filled epoxy sample 1.*

observation is that there is a peak shift towards the higher temperature side with increase in activated carbon powder content.

**Figure 4** shows the variation of DC conductivity with temperature for activated carbon powder epoxy gradient sample 3.

It has been observed that DC conductivity suddenly increases after 100°C in all the cases. Increase of DC conductivity appeared at 124°C and it goes on increasing up to 150°C.

**Figure 5** shows the variation of DC conductivity with temperature for activated carbon powder filled epoxy gradient composites sample 4. DC conductivity increases from 110°C then after 132°C there is an increase in DC conductivity.

It was reported that electrical conductivity of reinforced papers with respect to the weight fraction of Ag-plated carbon fiber increased with increasing content of carbon fiber. Due to the three-dimensional contacts between carbon fibers the electrical conductivity of the paper increased irrespective of the increase in thickness. The electrical conductivity of the reinforced paper having the Ag-plated carbon fiber was high because of the large number of pores formed on the activated carbon fiber [30]. When the volume percent of carbon content is increased or decreased, the material exhibits a change in resistivity. Heating can affect the conductivity of the polymer material on increasing the temperature [31]. On increasing

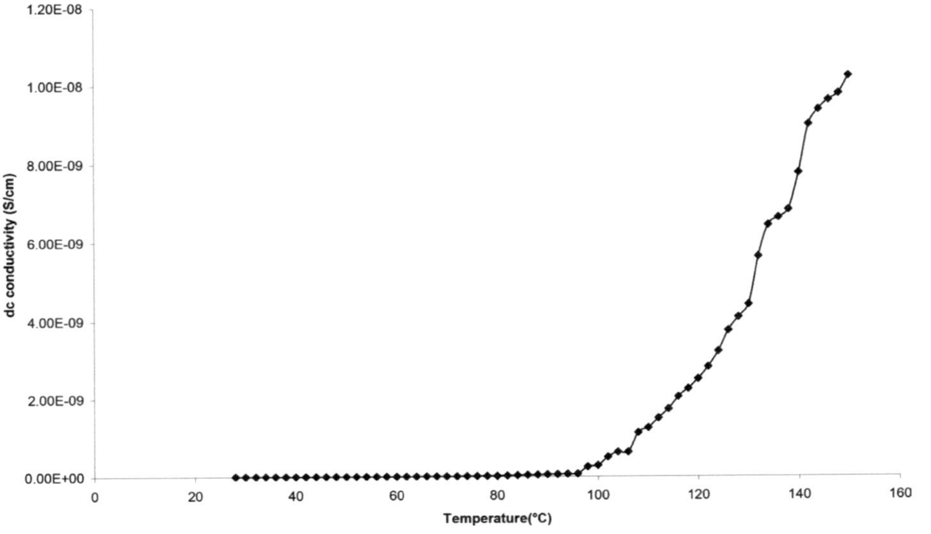


**Figure 3.**
*Variation of DC conductivity with temperature for activated carbon filled epoxy sample 2.*

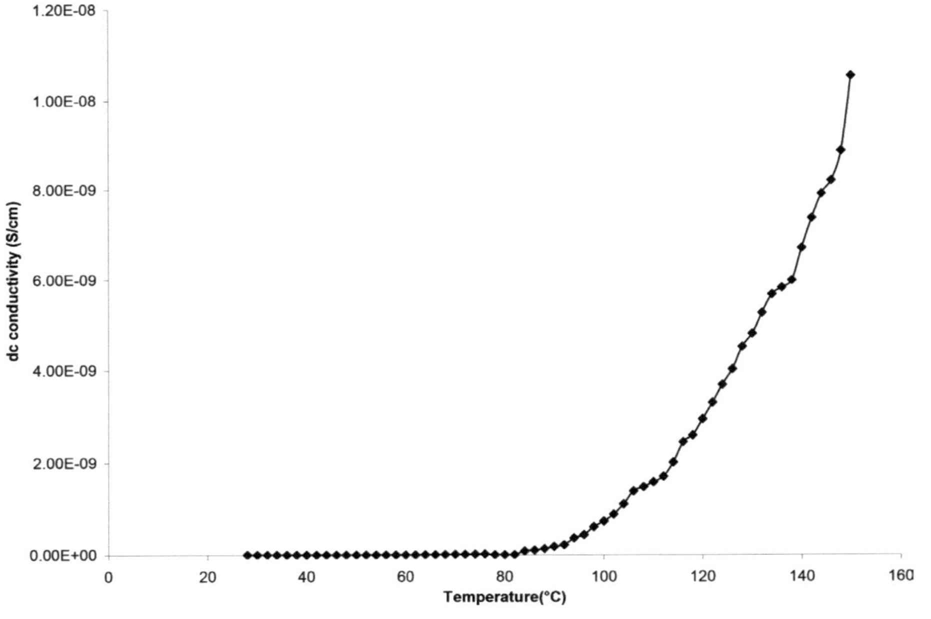


**Figure 4.**
*Variation of DC conductivity with temperature for activated carbon filled epoxy sample 3.*

temperature polymer expands as compared to CB aggregates and the interparticle distance between the aggregates increases, which causes destruction of conductive networks and as a result there is an increase in the resistivity with temperature.

It was reported that the electrical resistivity ($\rho$) of composite at low temperature is dominated by the electronic properties of the nanotubes, and tunneling nature [32–34].

$\ln \sigma$ vs. $T^{-1}$ plot for activated carbon powder filled epoxy has been analysed by using the following Arrhenius equation.

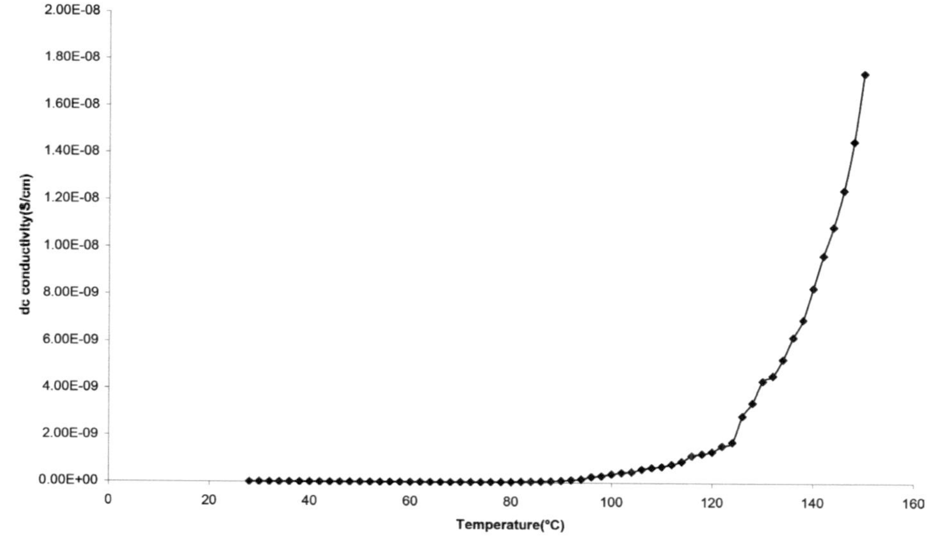


**Figure 5.**
*Variation of DC conductivity with temperature for activated carbon filled epoxy sample 4.*

| Sample no. | Activation energy (eV) |
|---|---|
| Sample 1 | 1.056682 |
| Sample 2 | 1.224812 |
| Sample 3 | 1.278476 |
| Sample 4 | 1.297912 |

**Table 2.**
*Lists the activation energy (eV) of sample 1, sample 2, sample 3, and sample 4.*

$$\sigma = A\ \exp^{-W_E/KT} \qquad (3)$$

where $W_E$ is the activation energy of conduction; k is Boltzmann's constant; A is a constant; and T is the temperature in (K).

On increasing filler concentration, conductive paths among the filler particles increase, and the average distance becomes smaller as a result conductivity of the composite increased.

## 4. Conclusions

a. DC conductivity value increases from sample 1 to sample 4. This shows the existence of graded structure (**Table 2**).

b. Increase of activated carbon content increases the DC conductivity.

c. Different transition points are observed in DC conductivity plots in different samples. Transition temperature shifts to lower side with the increase in activated carbon content.

d. Activation energy decrease with increase of activated carbon content in the samples.

e. These developed graded polymeric matrix composites having gradual variation of composition from carbonaceous filler at one surface to polymer dominated other end can be developed for desired electrical applications. Different type of sensors such as electrical resistance sensors, current sensors and temperature dependent sensors are required for various applications. Graded polymeric composites show variable resistivity behaviour, which can have potential applications in electromagnetic shielding, antistatic, corrosion-resistant coatings, conducting capabilities, light emitting devices, batteries and sensors. By virtue of the improved thermal stress relaxation and adhesive properties etc. Graded polymeric composites show variable resistivity behaviour, which can have potential applications in electromagnetic shielding, antistatic, corrosion-resistant coatings, conducting capabilities, light emitting devices, batteries and sensors.

## Author details

Archana Nigrawal
Advanced Materials and Processes Research Institute (AMPRI), (CSIR) (Formerly RRL Bhopal), Bhopal, India

*Address all correspondence to: archananigrawal@yahoo.co.in

## References

[1] Santos LF, Bianchi RF, Faria RM. Electrical properties of polymeric light-emitting diodes. Journal of Non-Crystalline Solids. 2004;**338-340**:590-594

[2] Kamimura S, Seguchi T, Ohamura K. Development of silicon nitride fiber from Si-containing polymer by radiation curing and its application. Radiation Physics and Chemistry. 1999;**54**:575-581

[3] Sevil UA, Guven O, Kovas A, Slezsak I. Gamma and electron dose response of electrical conductivity of polyaniline based polymer composites. Radiation Physics and Chemistry. 2003;**67**:575-580

[4] Chand N, Nigrawal A. Investigations on d.c. conductivity behaviour of milled carbon fibre reinforced epoxy graded composites. Bulletin of Materials Science. 2008;**31**(4):1-4

[5] Wang HL, Huang F, MacDiarmid AG, Wang YZ, Gebier DD, Epstein AJ. Application of polyaniline (emeraldine base, EB) in polymer light-emitting devices. Synthetic Metals. 1996;**78**:33-37

[6] Nguyen MT, Kasai P, Miller JL, Diaz AF. Synthesis and properties of novel water-soluble conducting polyaniline copolymers. Macromolecules. 1994;**27**:3625-3631

[7] Nigrawal A, Chand N. Effect of temperature on electrical and thermal properties on carbon soot filled polyester graded composites. Polymer-Plastics Technology and Engineering. 2011;**50**:547-551. ISSN: 03602559, 15256111

[8] Tsotra P, Friedrich K. Short carbon reinforced epoxy resin/polyaniline blends their electrical and mechanical properties. Composites Science and Technology. 2004;**64**:2385

[9] Wilson MS, Busick DN. Composite bipolar plate for electrochemical cells. US Patent No. 6,248,2001,467

[10] Hughes JDH. The carbon fibre/epoxy interface—A review. Composites Science and Technology. 1991;**41**:13-45

[11] Skotheim TA. Handbook of Conductive Polymers. Basel: Marcel Dekker; 1986

[12] Thostenson ET, Ren ZF, Chou TW. Advances in the science and technology of carbon nanotubes and their composites: A review. Composites Science and Technology. 2001;**61**:1899-1912

[13] Esawi AMK, Farag MM. Carbon nanotube reinforced composites: Potential and current challenges. Materials & Design. 2007;**28**:2394-2401

[14] Ajayan PM, Stephan O, Colliex C, Trauth D. Aligned carbon nanotube arrays formed by cutting a polymer resin–nanotube composite. Science. 1994;**256**:1212-1214

[15] Griebel M, Hamaekers J. Molecular dynamics simulations of the elastic moduli of polymer-carbon nanotube composites. Computer Methods in Applied Mechanics and Engineering. 2004;**193**:1773-1788

[16] Fidelus JD, Wiesel E, Gojny FH, Schulte K, Wagner HD. Thermo-mechanical properties of randomly oriented carbon/epoxy nanocomposites. Composites Part A. 2005;**36**:1555-1561

[17] Zhu R, Pan E, Roy AK. Molecular dynamics study of the stress-strain behavior of carbon-nanotube reinforced Epon 862 composites. Materials Science and Engineering A. 2007;**447**:51-57

[18] Ashrafi B, Hubert P. Modeling the elastic properties of carbon nanotube array/polymer composites. Composites Science and Technology 2006;**66**:387-396

[19] Han Y, Elliott J. Molecular dynamics simulations of the elastic properties of polymer/carbon nanotube composites. Computational Materials Science. 2007;**39**:315-323

[20] Albert KJ, Lewis NS, Schauer CL, Sotzing GA, Stitzel SE, Vaid TP, et al. Cross-reactive chemical sensor arrays. Chemical Reviews. 2000;**100**:2595-2626

[21] Freund MS, Lewis NS. A chemically diverse conduction polymer-based "electronic nose". Proceedings of the National Academy of Sciences of the United States of America. 1995;**92**:2652-2656

[22] Bartnikas R, Eichhorn RM. Electrical Properties of Solid Insulating Materials: Part A: Molecular Structure and Electrical Behavior. USA: ASTM Special Technical Publication; 1983

[23] Harun MH, Saion E, Kassim A, Hussain MY, Mustafa IS, Omer MAA. Temperature dependence of AC electrical conductivity of PVA-PPy-$FeCl_3$ composite polymer films. Malaysian Polymer Journal (MPJ). 2008;**3**(2):24-31

[24] Harun MH, Saion E, Kassim A, Mahmud E, Hussain MY, Mustafa IS. Dielectric properties of poly(vinyl alcohol)/polypyrrole composite polymer films. Journal for the Advancement of Science & Arts. 2009;**1**(1)

[25] Conductive Carbon Black in Plastics; Technical Report S-39; Boston, MA: Cabot Corp

[26] Chand N. SAR Hashmi Indian patent. 2003

[27] Hao D, Wei C. Dynamic characteristics analysis of bi-directional functionally graded Timoshenko beams. Composite Structures. 2016;**141**:253-263

[28] Simsek M. Bi-directional functionally graded materials (BDFGMs) for free and forced vibration of Timoshenko beams with various boundary conditions. Composite Structures. 2015;**133**:968-978

[29] Sumita M, Asai S, Miyadera N, Jojima E, Miyasaka K. Colloid and Polymer Science. 1986;**264**:212

[30] Jang J, Ryu SK. Journal of Materials Processing Technology. 2006;**180**:66-73

[31] Miyasaka K, Watanabe K, Jojima E, Aida H, Sumita M, Ishkiwa K. Journal of Materials Science. 1982;**17**:1610

[32] Badaire S, Pichot V, Zakri C, Poulin P, Launois P, Vavro J, et al. Correlation of properties with preferred orientation in coagulated and stretch-aligned single-wall carbon nanotubes. Journal of Applied Physics. 2004;**96**:7509-7513

[33] Abdul Khalil HPS, Jawaid M, Firoozian P, Alothman OY, Paridah MT, Zainudin ES. Flexural properties of activated carbon filled epoxy nanocomposites. The Malaysian Journal of Analytical Sciences, 2014;**18**(2):391-397

[34] Ebrahimi F, Nia NF. Thermo-mechanical vibration analysis of sandwich beams with functionally graded carbon nanotube-reinforced composite face sheets based on a higher-order shear deformation beam theory. Mechanics of Advanced Materials and Structures. 2017;**24**(10)

Chapter 4

# The Stochastic Finite Element in the Natural Frequency of Functionally Graded Material Beams

*Nguyen Van Thuan and Noh Hyuk Chun*

## Abstract

In this study, the stochastic finite element solution is given to obtain the variability in the natural frequency of functionally graded material (FGM) beam. The elastic modulus is assumed to vary in the thickness direction and the width of the beam to vary as well in the longitudinal direction following the exponential law. The random material properties of elastic modulus and mass density of the FGM beam are assumed to be one-dimensional homogeneous stochastic processes. The stochastic finite element analysis of FGM beam is performed in conjunction with Monte Carlo simulation (MCS) employing the spectral representation method for 16, the description of random processes of the random material properties under consideration. The response variability of the natural frequency due to random elastic modulus is evaluated for various states of randomness. Furthermore, the investigation on the effect of the correlation between random elastic modulus and random mass density on the response variability is addressed in detail as well.

**Keywords:** functionally graded materials, finite element method, FGM beam, Monte Carlo simulation

## 1. Introduction

Functionally graded materials (FGM) have received considerable attention in many engineering applications, since the theory of FGM was firstly introduced in 1984. In general, FGM is made from the volume fractions of two or more material components that have continuous variation of material properties from one surface to another [1]. Nowadays, FGM suits the specific demand in different engineering applications, especially for high temperature environment applications of heat exchanger tubes, thermal barrier coating for turbine blades, thermoelectric generators, furnace linings, electrically insulated metal ceramic joints, space/aerospace industries, automotive applications, biomedical area, etc.

The manufacturing of FGM with fully specified profile of material gradation, however, is very difficult causing significant variability in their mechanical and structural properties. Therefore, proper handling of the randomness in the material properties is required for accurate prediction of structural response for safe and reliable design. The stochastic analysis is a useful analytical tool to predict the response of structures with random material properties. In this direction, there is a reasonable body of recent research on the effect of uncertainties in material properties on the mechanical behavior of FGM. Investigators used stochastic simulation to study the effect of microstructural randomness on stress in FGM [2]. Ferrante et al. studied the effect of non-Gaussian porosity randomness on the response of functionally graded plate [3]. Yang et al. dealt with the stochastic bending response of moderately thick FGM plates [4, 5]. The effect of random material properties on post buckling response of FGM plate are presented in Lal et al. [6].

However, the above mentioned literatures are for the static analysis. To the best of author's knowledge, few limited works have been done on the eigen analysis of FGM structures involving randomness in system parameters. Certain efforts have been made in the past to predict the dynamic behavior of structures with randomness. In most of the studies conducted, investigators dealt with the free vibration of functionally graded laminates with random material properties using first-order perturbation technique (FOPT) incorporating mixed type and semi-analytical approach to derive the standard eigenvalue problem [7]. Some of these papers presented the stochastic finite element method (SFEM) to investigate the natural frequency of functionally graded plates based on the higher-order shear deformation theory (HSDT) utilizing first-order reliability method and second-order reliability method [8]. In most cases, Jagtap et al. [9] examined the stochastic nonlinear free vibration response of FGM plate using HSDT with von-Karman kinematic nonlinearity via direct iterative stochastic finite element method. Shegokar et al. investigated the stochastic finite element nonlinear free vibration analysis of FGM beam with random material properties due to thermo-piezoelectric loadings [10]. The above mentioned literatures investigated the free vibration and nonlinear behavior of FGM beam and plate. The material properties, such as Young's modulus, shear modulus, and Poisson's ratio of FGM, are modeled as independent random variables.

In this chapter, the stochastic finite element solution is suggested to obtain the variability in the natural frequency of functionally graded material (FGM) beam. The elastic modulus and width of the FGM beam are assumed to vary in thickness and longitudinal directions following the exponential law. The uncertain material properties, such as modulus of elasticity and mass density of the FGM beam, are considered to be a one-dimensional homogeneous stochastic process. The stochastic finite element analysis of FGM beam is performed using the spectral representation method for the description of randomness in conjunction with Monte Carlo simulation (MCS). The response variability of natural frequency due to random elastic modulus and mass density in FGM beam is given. Furthermore, the effect of correlation between the two random parameters is observed as well.

## 2. Theory formulation of FGM

An FGM is defined to be a material which has a continuous gradation through-the-thickness (h). One side of the material is typically ceramic and the other side is typically metal. A mixture of the two materials composes the through-the-thickness characteristics. Let us consider a functionally graded beam shown in **Figure 1**. The parameters of the model FGM beam are as follows: $L$ is the length of the beam, $h$ is the thickness of the beam, and $b$ is the width of the beam.

The elastic material properties vary through-the-beam thickness according to the volume fractions of the constituents using power law distribution (as shown in **Figure 2**).

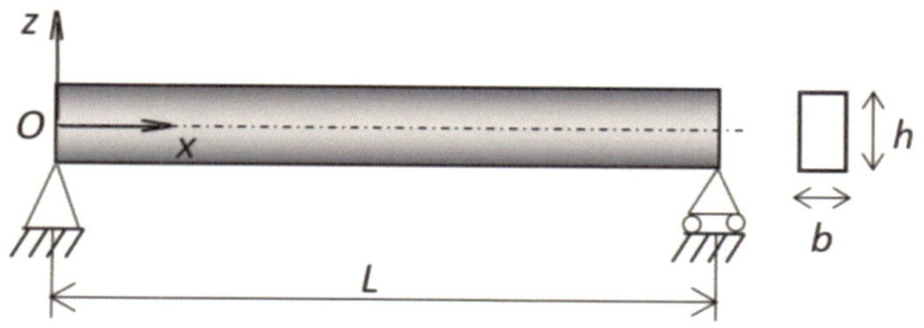


**Figure 1.**
*Model of FGM beams.*

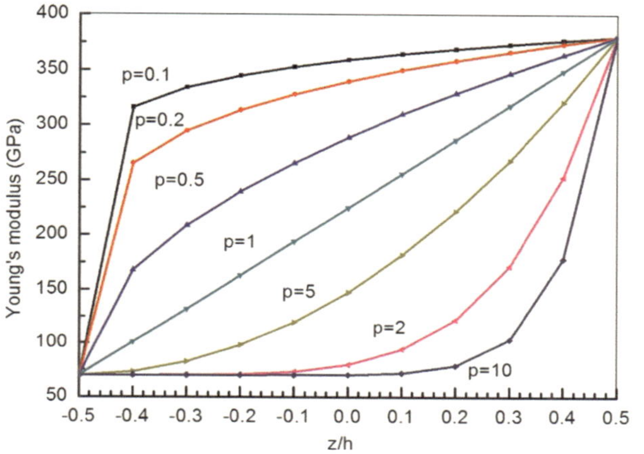


**Figure 2.**
*The variation of elasticity modulus along the corresponding directions.*

$$E(z) = (E_c - E_m)g(z) + Em$$

with

$$g(z) = \left(\frac{1}{2} + \frac{z}{h}\right)^p, \quad -0.5h \le z \le 0.5h, \quad 0 \le p \le \infty$$

where $E$ denotes the effective Young's modulus of elasticity, and $E_m$ and $E_c$ represent the Young's modulus of metal and ceramic, respectively. The parameters $g(z)$ and $p$ represent the volume fraction of the metal and ceramic exponent, respectively.

In exponential law, for the material parameter of Young's modulus $E$, mass density $\rho$ of the beam and the width of the beam $b$, with absolute values for $z$ coordinate, which endows the symmetric characteristic to the beam with respect to mid-plane.

$$E(z) = E_0 e^{\beta|z|}; \rho(z) = \rho_0 e^{\beta|z|}; b(x) = b_0 e^{\psi x}$$

In Eq. (3), $E_0$, $\rho_0$ are the values of the Young's modulus and mass density at the mid-plane ($z = 0$) of the beam. The parameter $\beta$ in the exponent characterizes the material property variation along the thickness direction. The width of the beam $b$ varies according to the non-uniformity parameter $\psi$ along the axis of the beam. In order to give the insight into the varying characteristics of the FGM beam under

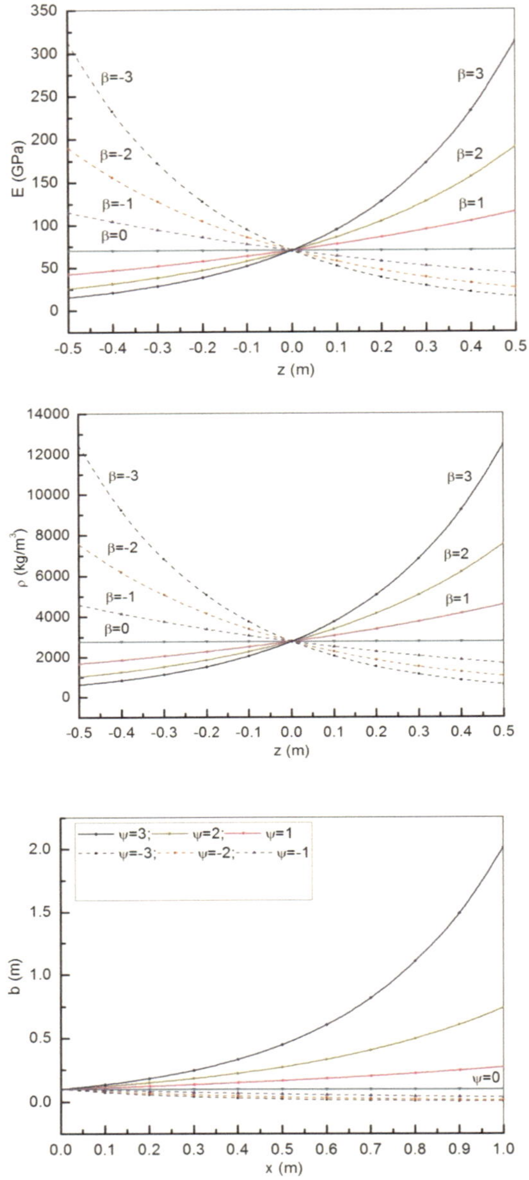


**Figure 3.**
*The variation of elasticity modulus, mass density, and width along the corresponding directions.*

consideration, change of material constants and width depending on the corresponding parameters are shown in **Figure 3**.

## 3. Finite element formulation for FGM beam and frequency analysis

### 3.1 Finite elements

In case of four degree of freedom beam element, as shown in **Figure 4**, the transverse displacement function may be assumed as a cubic polynomial in $x$, and the corresponding shape functions are Hermite interpolation functions.

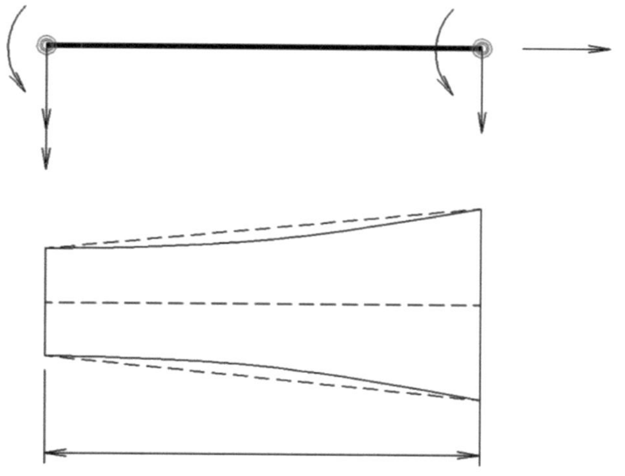

**Figure 4.**
*Beam element.*

The width and mass per unit length of the element are

$$b_e(x) = b_1 + \frac{b_2 - b_1}{L_e}x,$$
$$m_e(x) = mb_e(x) = m\left(b_1 + \frac{b_2 - b_1}{L_e}x\right),$$

where $m = \int_{-h/2}^{h/2} \rho(z)dz$ and $\rho(z)$ denote the mass density at $z$.

The nodal displacement vector of the element is

$$\{q\}_e = \{q_1 \; q_2 \; q_3 \; q_4\}^T,$$

then, the displacement field is

$$w_e = \langle N \rangle \{q\}_e,$$

where $\langle N \rangle = \langle N_1 \quad N_2 \quad N_3 \quad N_4 \rangle$, and $N_i$ is Hermite shape function of $i$-th degree of freedom.

In this case, the stiffness of the beam $b_e D_{11}$ is similar to $EI$ of the homogeneous beam

$$D_{11} = \int_{-h/2}^{h/2} \frac{E(z)}{1 - \nu^2} z^2 dz,$$

where $\nu$ is Poisson's ratio.

## 3.2 Application of Hamilton's principle

Hamilton's principle may be a theoretical base for dynamical systems by its nature of integral form in time with Lagrangian density to account for continuous space. In this paper, the analysis of natural frequency of FGM beam is performed using Hamilton's principle. The strain energy expression $U_e$ for bending is given as following:

$$U_e = \int_0^{L_e} \frac{b_e D_{11}}{2} \left( \frac{\partial^2 w_e}{\partial x^2} \right)^2 dx.$$

The kinematic energy $T_e$ for flexural vibration is

$$T_e = \int_0^{L_e} \frac{m b_e \dot{w}_e^2}{2} dx.$$

where

$$K_e = \int_0^{L_e} b_e D_{11} \langle N \rangle^T \langle N \rangle dx.$$

$$M_e = \int_0^{L_e} m_e b_e \langle N \rangle^T \langle N \rangle dx.$$

Substituting Eq. (7) into Eqs. (9) and (10), the following can be obtained:

$$U_e = \frac{1}{2} \{q\}_e^T K_e \{q\}_e,$$
$$T_e = \frac{1}{2} \{\dot{q}\}_e^T M_e \{\dot{q}\}_e.$$

The governing differential equations of motion and the related governing equation can be derived using Hamilton's principle

$$\delta \int_{t_1}^{t_2} \left( \sum_1^N U_e - \sum_1^N T_e \right) dt = 0,$$

where $N$ denotes the number of finite elements.

Substituting Eqs. (13) and (14) into Eq. (15), the following can be obtained:

$$M\ddot{q} + Kq = 0,$$

here, $q$ is the nodal displacement of the beam.

For simple harmonic vibration, we assume the displacements to be $q(x,t) = w(x)e^{i\omega t}$. Accordingly with Eq. (16), we can obtain the following:

$$(K - \omega^2 M)w = 0,$$

where $K$ is the assembled global stiffness matrix of $K_e$, the element stiffness matrix, which is given in detail in the Appendix. For Eq. (17) to be valid, i.e., to have nontrivial solution, the following needs to be satisfied:

$$\det(K - \omega^2 M) = 0.$$

## 4. Modeling of randomness

### 4.1 Mathematical expression

In order to model the randomness in the material properties, the modulus of elasticity and mass density along the mid-plane are assumed to vary along its length of FGM beam in a random manner. We can model these variations as one-dimensional univariate (1D-1V) homogeneous stochastic processes. The simple mathematical expressions for the randomly varying modulus of elasticity and mass density can be written as

$$E_0(x) = \overline{E}_0\left[1 + f_E(x)\right],$$
$$\rho_0(x) = \overline{\rho}_0\left[1 + f_\rho(x)\right],$$

where $x$ is the coordinate along the axis of the FGM beam, $\overline{E}_0, \overline{\rho}_0$ are the expected values of $E_0$ and $\rho_0$, respectively, and $f_E(x), f_\rho(x)$ are one-dimensional stochastic process which are homogeneous with zero-mean values.

The numerical generation of sample functions of Gaussian zero-mean homogeneous stochastic processes, which describe the randomness in parameters of the structure, is accomplished using the spectral representation method. For a one-dimensional univariate (1D-1V) stochastic process, we have [11]

$$f(x) = \sqrt{2}\sum_{n=0}^{N-1} A_n \cos\left(\omega_n t + \phi_n\right),$$
$$A_n = \sqrt{2S_{ff}(\omega_n)\Delta\omega},$$
$$\Delta\omega = \frac{\omega_u}{N},$$
$$\omega_n = n\Delta\omega, n = 0, 1, 2, \ldots, N-1.$$

In Eq. (21), $\omega_u$ denotes the upper cut-off frequency beyond which the power spectral density function $S_{ff}(\omega_n)$ may be assumed to be zero for either mathematical or physical reasons. The following criterion is usually used to estimate the value of $\omega_u$:

$$\int_0^{\omega_u} S_{ff}(\omega_n)d\omega_n = (1-\varepsilon)\int_0^{\infty} S_{ff}(\omega_n)d\omega_n,$$

where $\varepsilon << 1$.

The uniform random phase angle $\phi_n$ in Eq. lies in the range of $[0, 2\pi]$. The power spectral density function used in Eq. (20) is given as

$$S_{ff}(\omega_n) = \frac{1}{\sqrt{\pi}}\sigma_f{}^2 d\, e^{\left(-d^2\omega_n^2\right)}, \quad -\infty < \omega_n < \infty,$$

here, $\sigma_f$ denotes the standard deviation of the stochastic process $f(x)$ and $d$ is the correlation distance of the stochastic process along the $x$ axis.

In all examples, the coefficient of variation $(COV)$ of natural frequency, which is defined as a ratio of the standard deviation of response to the absolute mean response, will be used to give the variability of the response.

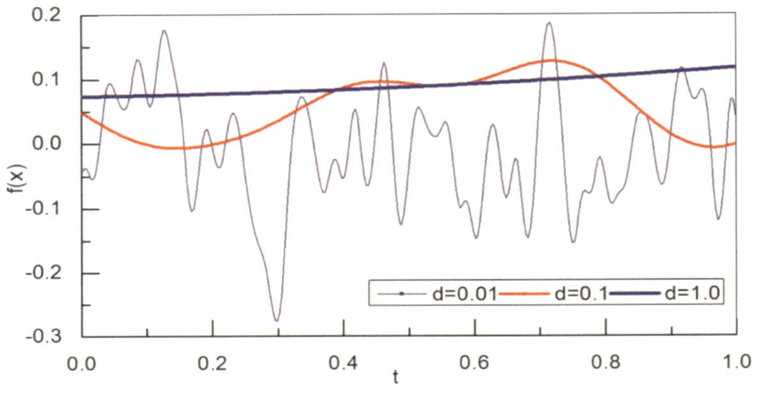


**Figure 5.**
*Example plot of random process used* $(d = 0.01, 0.1, 1.0, \sigma_f = 0.1)$.

$$COV = \frac{\text{Standard deviation}}{|Mean|},$$

here,

$$\left.\begin{array}{l} \text{Standard deviation} = \sqrt{\text{var}(\omega)} \\ Var(\omega) = E\left[(\omega - \overline{\omega})^2\right] \end{array}\right\}.$$

and $\omega$ is the natural frequency of the FGM beam, and $\overline{\omega}$ denotes the mean of the natural frequency.

### 4.2 Monte Carlo analysis

In order to obtain the response variability in the natural frequency of the FGM beam, we employed the scheme of Monte Carlo simulation (MCS). As a matter of fact, the MCS corresponds to the deterministic analyses on a set of heterogeneous models of the given structure, in which the material properties have different values depending on the position in the domain of the structure.

The generation of heterogeneous random samples is accomplished by the aforementioned spectral representation scheme, and we use 10,000 samples for respective analyses. In particular, we adopt the local average scheme other than the midpoint rule in applying the MCS, with which better results can be obtained especially for the processes with small correlation distance. **Figure 5** shows an example plot of the processes employed to model the randomness in the system parameters.

## 5. Numerical example

The geometric dimensions of the example FGM beam are: $h$ = 0.1 m, $L$ = 1 m, and $b_0$ = 0.1 m. It is assumed that the material properties are $E_0$ = 70 GPa, $\rho_0$ = 2780 kg/m$^3$, and $\nu$ = 0.33. $E_0$ denotes the Young's modulus at the mid-surface of the beam and $E_1$ at the top and bottom surfaces following Eq. (3) (**Figure 6**).

### 5.1 Deterministic analysis results

The results in **Figure 7** correspond to the prismatic homogeneous beam since the parameters in exponents, $\beta$ and $\psi$, are all zero. The natural frequency is obtained in

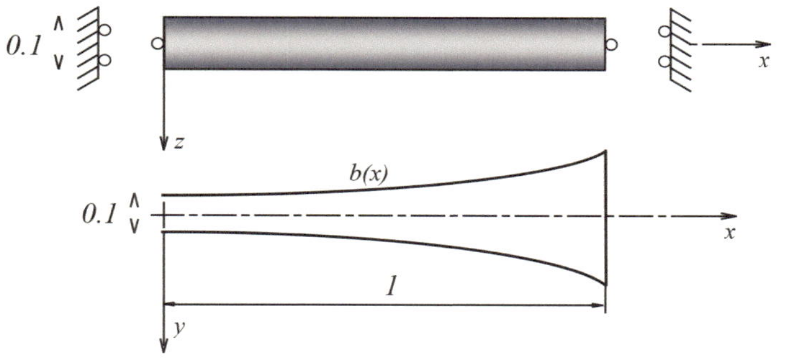


**Figure 6.**
*FGM beam model.*

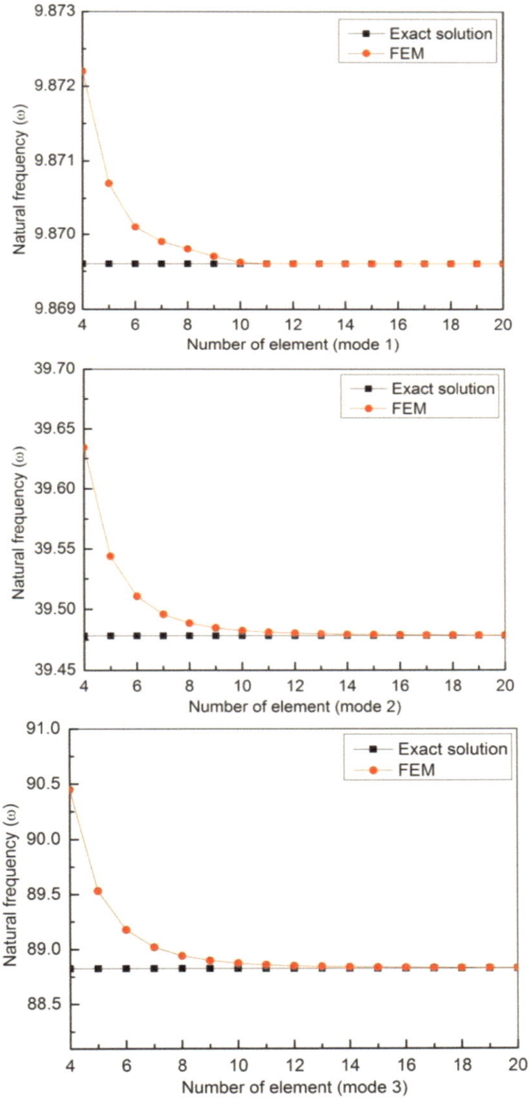


**Figure 7.**
*The convergence between exact solution and finite element method.*

the present study to compare with the results of analytical solution given by Hassen et al. [12].

The discrepancies between exact and finite element solutions for the frequencies for the first three modes are shown in **Figure 8**. The differences given in percentile tends to zero as the number of finite elements is increased, meaning the results are converging to exact solutions.

**Figure 9** shows the first three normalized natural frequencies of uniform FGM beams for three cases of modulus ratio ($E_1/E_0$). The natural frequency increases as the ratio of Young's modulus increases from $E_1 = 0.2E_0$ to $E_1 = 5E_0$.

## 5.2 Variability of natural frequency due to randomness in elastic modulus

**Figure 10** shows the COV of response versus the correlation distance ($d$) when the elastic modulus $E_0$ is random. The standard deviation of the random elastic modulus is denoted by $\sigma_f$. In all cases, the COV of natural frequency shows similar trends, starting from small values for small correlation distance, up to large values

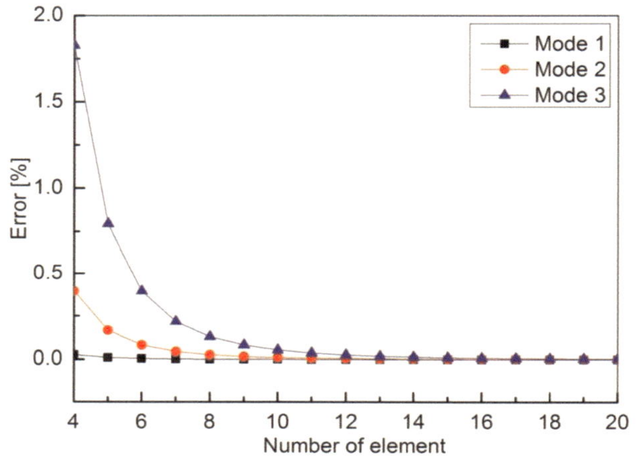


**Figure 8.**
*Error depending on mesh refinement ($\beta = 0, \psi = 0$).*

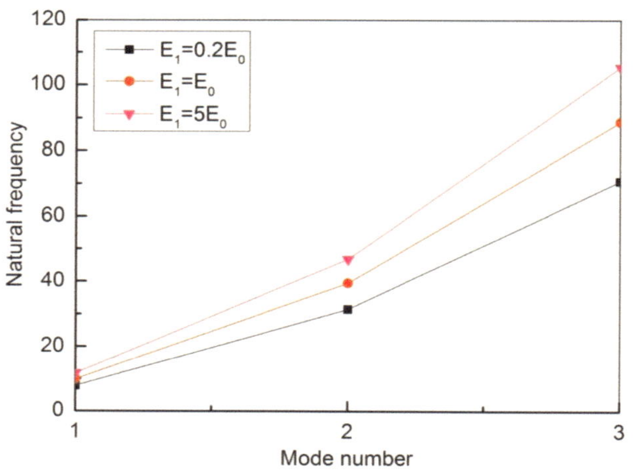


**Figure 9.**
*First three normalized natural frequencies of FGM beams for different ratio of Young's modulus ($\psi = 0$).*

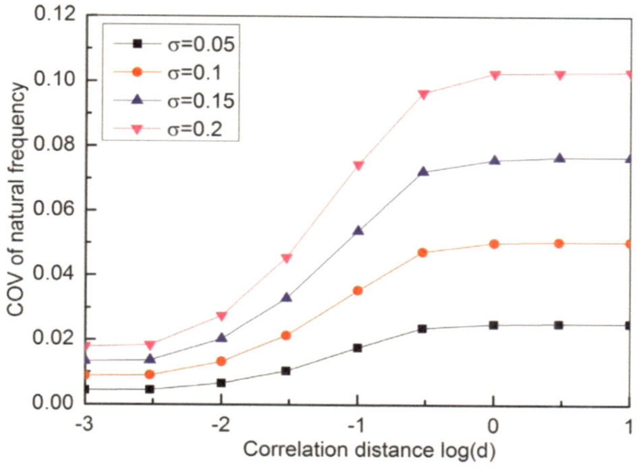


**Figure 10.**
*COV of natural frequency as a function of the correlation distance (d) for different standard deviation of stochastic process.*

for large correlation distance. In the FGM beam under consideration, the response variability, when the correlation distance tends to infinity, is obtained to be about 50% of the input standard deviation of the stochastic process.

The relationship between COV of natural frequency and the COV of stochastic process is shown in **Figure 11**. The standard deviation of stochastic process is changed from 0.0 to 0.25. As seen in **Figure 11**, the COV of response shows a slightly nonlinear pattern in all the cases of $d = 0.01$, 0.1, and 10.

The effect of mesh refinement on the COV of natural frequency is shown in **Figure 12**. The correlation distance log $(d)$ is assumed to be from $-3$ to 1. As seen in **Figure 12**, the COV of natural frequency is not affected by the mesh refinement.

**Figure 13** shows the variation of the coefficient of variation (COV) depending on the non-uniformity parameter $(\psi)$ of the FGM beam. As seen in the figure, the COV of response is not slightly affected by the non-uniformity parameter in particular for large correlation distances.

The overall features of the effect of Young's modulus ratio on COV of natural frequency are shown in **Figure 14**. The COV of natural frequency is not affected by the parameter $\beta$. These results can easily be understood because the standard deviation and the mean of natural frequency in Eq. (27) increase in the same rate.

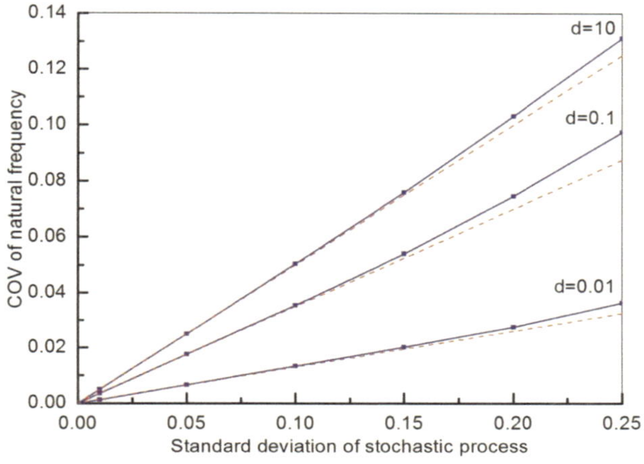


**Figure 11.**
*COV of natural frequency as a function of the standard deviation of stochastic process* $f_E(x)$.

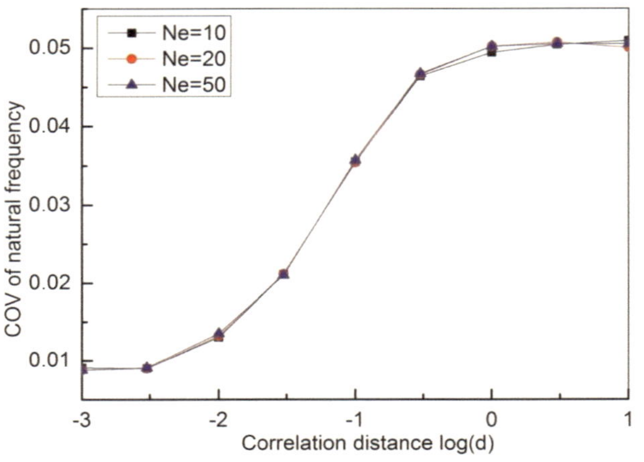


**Figure 12.**
*Effect of mesh refinement on the COV of natural frequency ($\sigma_f = 0.1$).*

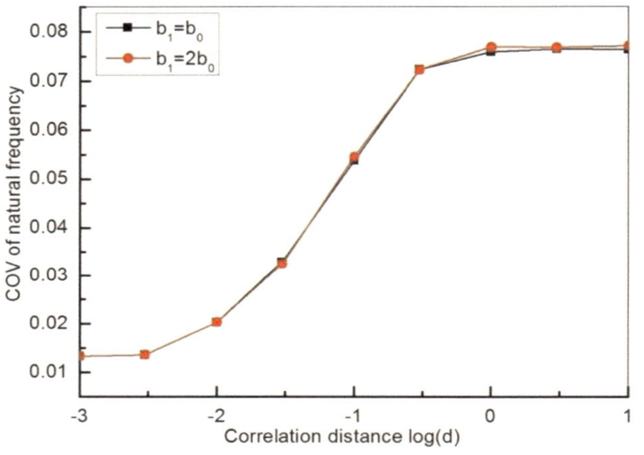


**Figure 13.**
*Effect of non-uniformity parameter ($\psi$) on COV of natural frequency ($Ne = 20$, $\sigma_f = 0.15$).*

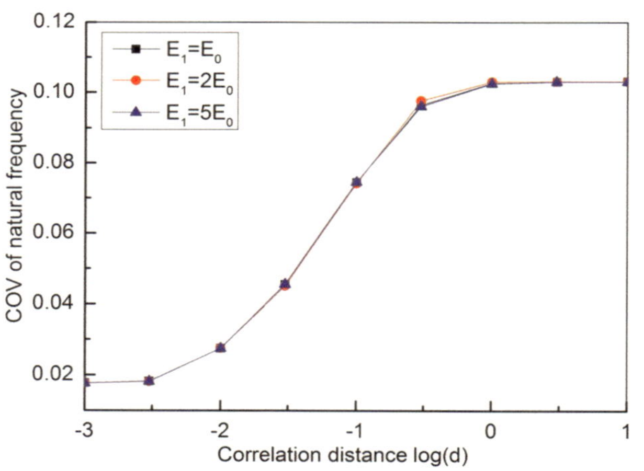


**Figure 14.**
*Effect of parameter $\beta$ on the COV of natural frequency ($Ne = 20$, $\sigma_f = 0.2$).*

### 5.3 FGM beam having correlation multiple randomness

It is natural to have preposition that not only the elastic modulus, but also the mass density of the material can have randomness. Therefore, we need to consider

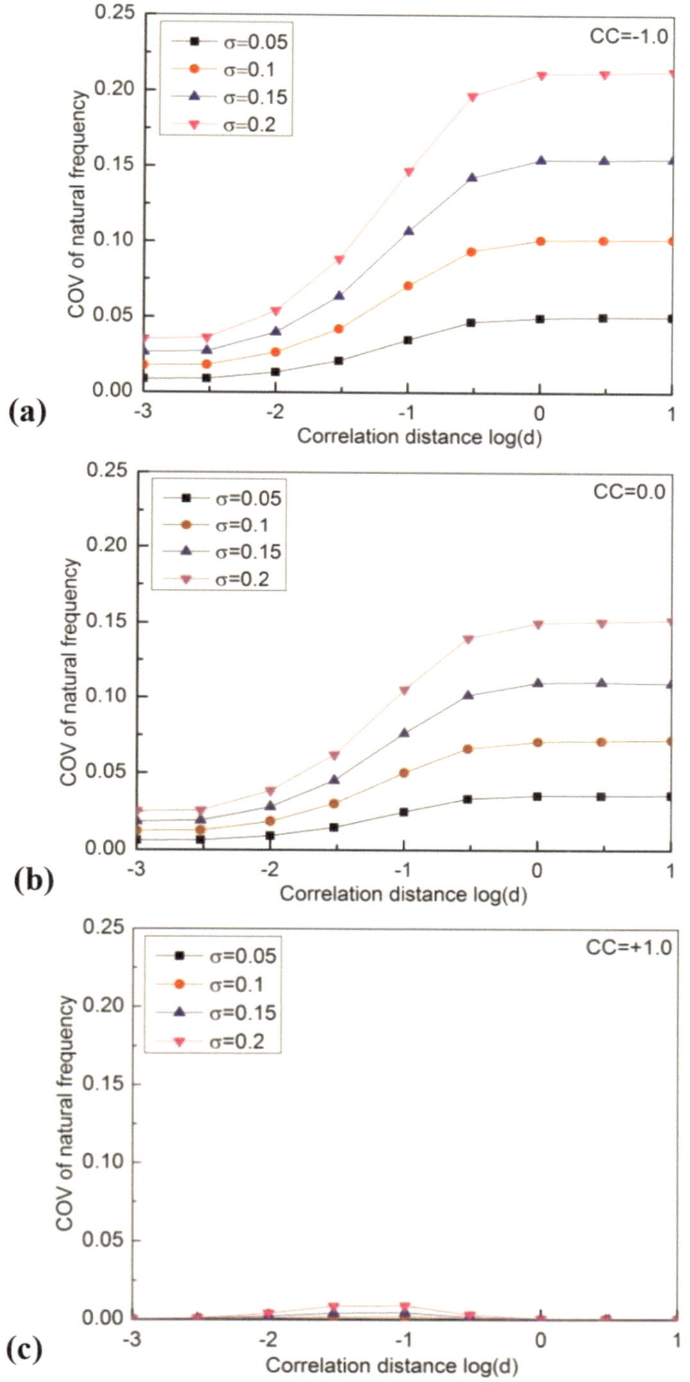


**Figure 15.**
*Effect of the correlation between two random parameters: (a) negative perfect correlation (CC = −1.0), (b) no correlation (CC = 0.0), and (c) positive perfect correlation (CC = +1.0).*

the effect of correlation between two random parameters of elastic modulus and mass density. To this aim, we consider three correlation cases: +1.0, 0.0, and −1.0. When the random processes for elastic modulus, $f_E$, and mass density, $f_\rho$, are exactly the same, the correlation coefficient (CC) is +1.0. If the values have negative

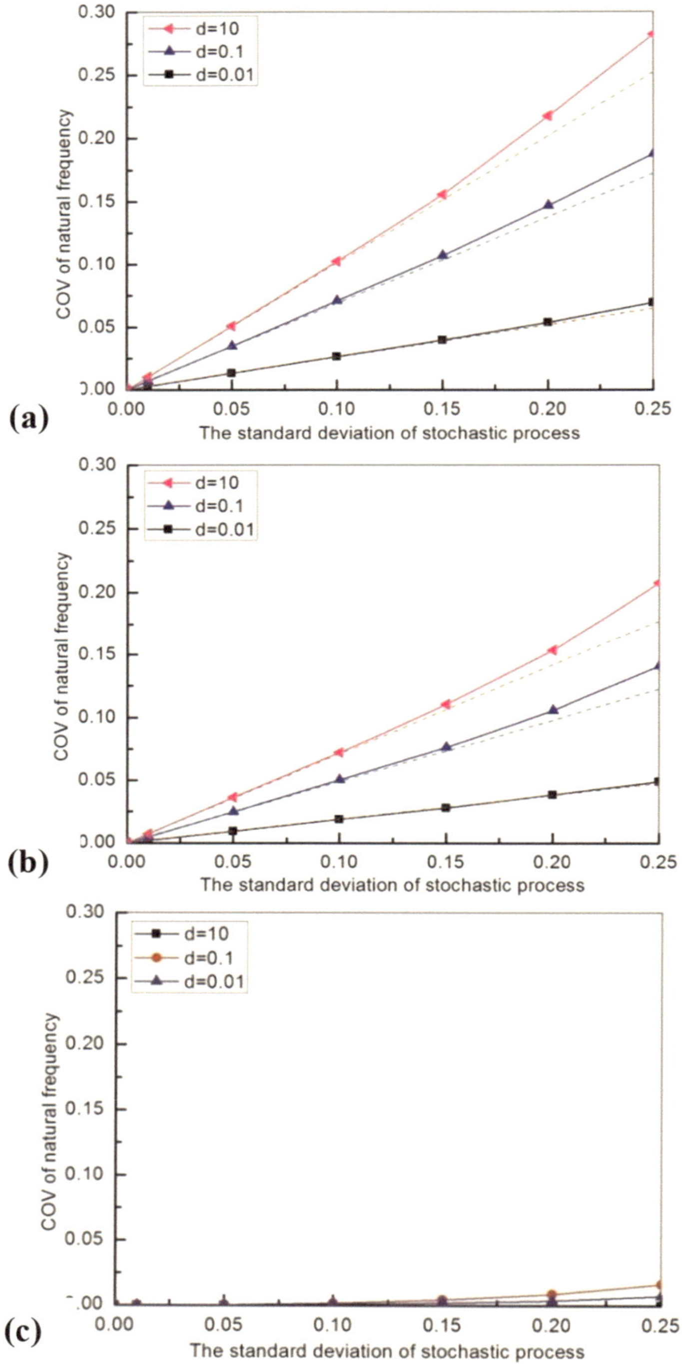


**Figure 16.**
*COV of natural frequency as a function of the standard deviation of stochastic process: (a) negative perfect correlation (CC = −1.0), (b) no correlation (CC = 0.0), and (c) positive perfect correlation (CC = +1.0).*

amounts, then the correlation coefficient is −1.0. The zero-correlation (CC = 0.0) means that stochastic processes of the two parameters are theoretically independent [13].

As shown in **Figure 15**, the maximum COV exceeds the input standard deviation of the stochastic process in the case of negative perfect correlation, while it is about 75% in the case of no correlation. However, in case of positive perfect correlation (**Figure 15c**), the response COV of natural frequency of FGM beam is small enough to be ignored. These results can easily be understood because the stiffness and mass matrix of each elements in Eq. (18) increase or decrease with the same rate in the case of positive perfect correlation. In case of negative perfect correlation, the ratio of random parts in the stiffness and mass are relatively large since the random parts have opposite sign, which makes the response variability large.

In the case, when we take the correlated multiple random material properties into account, we also obtained the slight nonlinear pattern of variation of COV in terms of COV of stochastic fields, as shown in **Figure 16**.

## 6. Conclusions

To evaluate response variability due to a single parameter of the random Young's modulus and multiple uncertain material properties, a formulation in the context of stochastic finite element solution is suggested for the natural frequency of FGM beam. In deriving the formula for the covariance of the response, modified power spectral density and correlation function are defined by using the general formula of random processes. The Monte Carlo simulation is performed employing the statistical preconditioning scheme as a random process generation technique. The local average method is employed instead of mid-point rule in Monte Carlo simulation.

In FGM beam natural frequency, the response COV for correlation between a random of single parameter and two uncertain parameters is observed. The coefficient of variation of natural frequency can only reach about 50% of the input standard deviation of the stochastic process in a single parameter of the random elastic modulus. However, the number of values is increased over 100% of the input standard deviation of the stochastic process in multiple uncertain material properties, when the correlation distance tends to infinity. The results showed that the COV of natural frequency of FGM beam in a single parameter of the random Young's modulus and multiple uncertain material properties achieved maximum variability for $d$ about 1.0.

There is a very small difference between deterministic natural frequency and probabilistic natural frequency of FGM beam for the case of positive perfect correlation. Also, the COV of natural frequency does not depend on the number of elements, Young's modulus ratio, and the ratio of non-uniformity parameter of FGM beam. The importance of these parameters needs to be studied as a further work.

## A. Appendix

Hermite shape functions of beam finite element:

$$N_1 = x\left(1 - 2\frac{x}{L} + \frac{x^2}{L^2}\right); N_2 = 1 - 3\frac{x^2}{L^2} + 2\frac{x^3}{L^3}; N_3 = x\left(-\frac{x}{L} + \frac{x^2}{L^2}\right); N_4 = 3\frac{x^2}{L^2} - 2\frac{x^3}{L^3}.$$

Stiffness matrix:

$$K_e = \frac{D_{11}}{L^3}\begin{bmatrix} 6(b_2+b_1) & 2(b_2+2b_1)L & -6(b_2+b_1) & 2(2b_2+b_1)L \\ & (b_2+3b_1)L^2 & -2(b_2+2b_1)L & (b_2+b_1)L^2 \\ & & 6(b_2+b_1) & -2(2b_2+b_1)L \\ Sym. & & & (3b_2+b_1)L^2 \end{bmatrix}.$$

Mass matrix:

$$M_e = \frac{mL}{840}\begin{bmatrix} 24(3b_2+10b_1) & 2(b_2+b_1)L & 54(b_2+b_1) & -2(6b_2+7b_1)L \\ & (3b_2+5b_1)L^2 & 2(7b_2+6b_1)L & -3(b_2+b_1)L^2 \\ & & 24(10b_2+3b_1) & -2(15b_2+7b_1)L \\ Sym. & & & (5b_2+3b_1)L^2 \end{bmatrix}.$$

## Author details

Nguyen Van Thuan[1*] and Noh Hyuk Chun[2]

1 Nha Trang University, Vietnam

2 Sejong University, South Korea

*Address all correspondence to: thuannv@ntu.edu.vn

## References

[1] Koizumi M. FGM activities in Japan. Composites Part B Engineering. 1997; **28B**:1-4

[2] Ferrante FJ, Graham L. Randomness effects on the properties of a functionally graded plate using probabilistic methods. In: Structural Safety and Reliability: ICOSSAR '01; Newport Beach, USA; 2001

[3] Graham-Brady LL. Stochastic simulation of non-Gaussian/non-stationary properties in a functionally graded plate. Computer Methods in Applied Mechanics and Engineering. 2005;**194**:1675-1692

[4] Yang J, Liewb KM, Kitipornchaia S. Stochastic analysis of compositionally graded plates with system randomness under static loading. International Journal of Mechanical Sciences. 2005;**47**: 1519-1541

[5] Yang J, Liewb KM, Kitipornchaia S. Second-order statistics of the elastic buckling of functionally graded rectangular plates. Composites Science and Technology. 2005;**65**:1165-1175

[6] Lal A, Jagtap KR, Singh BN. Post buckling response of functionally graded materials plate subjected to mechanical and thermal loadings with random material properties. Applied Mathematical Modelling. 2013;**37**: 2900-2920

[7] Kitipornchaia S, Yang J, Liewb KM. Random vibration of the functionally graded laminates in thermal environments. Computer Methods in Applied Mechanics and Engineering. 2006;**195**:1075-1095

[8] Shaker A, Abdelrahman W, Tawfik M, Sadek E. Stochastic finite element analysis of the free vibration of functionally graded material plates. Computational Mechanics. 2008;**41**: 707-714

[9] Jagtap KR, Lal A, Singh BN. Stochastic nonlinear free vibration analysis of elastically supported functionally graded materials plate with system randomness in thermal environment. Composite Structures. 2011;**93**:3185-3199

[10] Shegokar NL, Lal A. Stochastic finite element nonlinear free vibration analysis of piezoelectric functionally graded materials beam subjected to thermo-piezoelectric loadings with material uncertainties. Meccanica. 2014; **49**:1039-1068

[11] Shinozuka M, Deidatis G. Simulation of stochastic processes by spectral representation. Applied Mechanics Reviews. 1991;**44**(1):191-201

[12] Haasen AA, Abdelouahed T, Sid AM, Hichem AB. Free vibration behavior of exponential functionally graded beams with varying cross-section. Journal of Vibration and Control. 2011;**17**(2):311-318

[13] Noh HC. Effect of multiple uncertain material properties on the response variability of in-plane and plate structures. Computer Methods in Applied Mechanics and Engineering. 2006:2697-2718

Chapter 5

# Development of Functionally Gradient Cu-Sn-Ni Alloy Using GTA Heat Source

*Cherian Paul and Ramasamy Sellamuthu*

## Abstract

The impact of nickel content on surface hardness, microstructure and wear properties of surface alloyed Cu-10Sn bronze composite was examined in this chapter. Gas Tungsten Arc (GTA) was utilized as the heat source for the surface alloying/modification process. The surface modification process was carried out on bronze samples coated with various Nickel coating thicknesses. Vickers hardness tester was used to measure the surface hardness as well as the hardness along the depth of the modified layer and wear rate was measured using a pin-on-disc tribometer. The Ni concentration profiling was carried out using EDAX. Surface modification process resulted in the formation of a layered functionally graded bronze alloy. The average grain size was found to reduce upon surface modification process. Ni addition was observed to increase the hardness and reduce wear rate for the modified samples.

**Keywords:** nickel profile, heat source, functionally graded material, FGM, hardness, wear behavior, surface modification

## 1. Introduction

Bronze, owing to its superior wear resistance is generally treated as one of the most commonly used engineering materials mainly as a bearing material in aerospace, automotive as well as industrial applications. Researches are being conducted on application of traditional coating methods like PVD, CVD, sputter deposition, electroplating, etc., for improving the surface properties of bronze. Surface modification process (SMP) has become an emerging technique to replace the traditional coating processes to improve the tribological properties of ferrous as well as non-ferrous alloys. In SMP, a heat source is used to melt the substrate surface and thereby a molten pool is formed. Then, the heat source is progressively moved along the length of the substrate so that, upon solidification a modified layer will be formed. In the case of fixed heat source, substrate will be moved. The major advantage of using SMP is that, the modified layer formed after solidification is integral to the substrate Benkisser et al. [1]. The applications of the alloys can be extended to ship propellers, sub-sea weapon ejection system, pumps, bearings and bushes as well. The drawback of traditional coatings getting delaminated on repeated cycles of operation can be omitted by using SMP. The formation of a functionally graded material (FGM) can be expected as a result of SMP. Since FGM is characterized by the gradual variation in composition and structure over volume, SMP with alloying elements results in the formation of an FGM. Wear

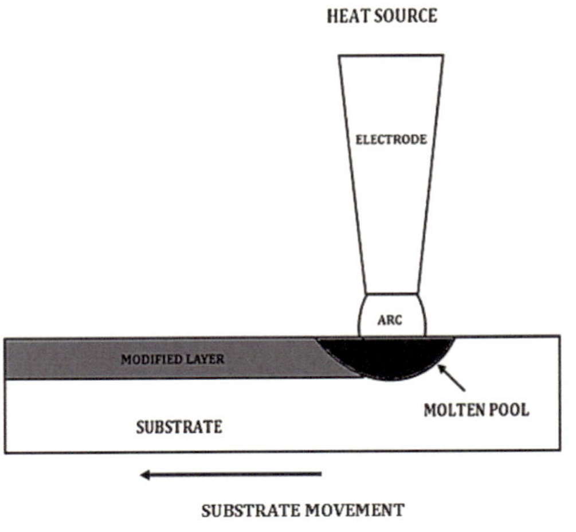


**Figure 1.**
*Surface modification process.*

| Process variable | Value | Unit |
|---|---|---|
| Current | 200 | A |
| Electrode diameter | 2.4 | mm |
| Arc length | 1.5 | mm |
| Electrode angle | 180 | ° |
| Traverse speed | 1 | mm/s |
| Argon flow rate | 12 | l/min |

**Table 1.**
*GTA process variables.*

**Figure 2.**
*Experimental setup.*

resistance, hardness, corrosion resistance, thermal conductivity, etc., of various alloys can be improved by the application of SMP. C Paul [2]. A schematic of SMP is shown in **Figure 1**.

This chapter discusses an investigation on the effect of Ni content on the hardness, the wear rate and the coefficient of friction of the surface alloyed Cu-Sn bronze alloy. As no previous works have been reported in the literature on the effect of Ni content on the hardness and the wear behavior of the surface alloyed bronze alloy, the present research work is undertaken. In the present study, the Sn content of the alloy was kept constant at 10 wt % and the Ni content was varied. The bronze substrates are coated with Ni of varying coating thickness (80, 120, 160 and 200 μm) using electroplating technique. The surface alloying process was carried out on the Cu-Sn bronze alloy coated with Ni. The GTA was used as the heat source. The GTA process variables, current (I), electrode diameter (eΦ), arc length (l), electrode angle (eθ), traverse speed (u) and argon flow rate are kept constant during the surface alloying process. The GTA process variables used in this study are reported in **Table 1**. The Ni concentration profiling was carried out for the surface alloyed samples. The Ni concentration was measured on the surface as well as along the depth of the modified layer formed in the surface alloying process using EDAX

**Figure 3.**
*Cu-Sn alloy with and without Ni coating.*

**Figure 4.**
*Cu-Sn alloy surface modified with Ni.*

analysis. Microstructural examination, hardness measurement and wear measurements were carried out for the substrate and for the specimen surface alloyed with Ni. **Figure 2** shows the experimental setup.

**Figure 3** shows the Cu-Sn alloy substrate with and without Ni coating and **Figure 4** shows the surface modified Cu-Sn-Ni alloy.

## 2. Microstructural examination

A typical dendritic structure was observed in the as-cast substrate of Cu-10Sn bronze alloy as shown in **Figure 5** and the microstructure of the Ni surface alloyed Cu-10Sn alloy is shown in **Figure 6**.

It can be observed from **Figure 6** that the structure is very fine as opposed to a coarse structure observed in **Figure 5** and therefore it can be concluded that the grain refinement occurs as a result of the surface alloying process [3]. This refinement is due to the fast cooling experienced during solidification in the surface alloying process. A similar fine grained microstructure was observed for all the other Ni alloyed specimens also. Yilbas et al. [4, 5] studied the effect of laser surface modification treatment of aluminum bronze (Cu-9%Al-3%Fe) with

**Figure 5.**
*As-cast Cu-Sn.*

**Figure 6.**
*Ni surface alloyed Cu-Sn.*

B4C and reported that fine grains are formed at the laser treated surface in the surface modification process because of high cooling rate. Kac et al. [6] studied the structure and properties of surface alloyed aluminum bronze (Cu-10%Al-4%Fe-2%Mn) with Ti as the alloying element using laser heat source. They reported that a very fine microstructure was formed in the rapid solidification experienced in the laser process. Viswanadham et al. [7] studied the injection of TiC particles into aluminum bronze (Cu-7%Al-3%Fe-1.5%Mn) using the laser as the heat source. They have reported that the modified layer in the laser treated specimen was found to be dense and highly uniform when compared to the untreated specimen. Majumdar and Manna [8] carried out the surface alloying of pure Cu with Cr using the laser as the heat source and they have evaluated the microstructure resulting from the surface alloying process. They reported that the microstructure of the alloyed zone changed from coarse dendritic for the substrate to a fine dendritic structure in the surface alloying process. It can be concluded that the result obtained in the present study is consistent with that of the previous studies.

## 3. Ni concentration profile

The Ni concentration on the surface of the modified layer formed in the surface alloying process was measured using the EDAX analysis. The concentration along the depth of the modified layer was also measured. **Figure 7** shows the points where the Ni concentration was measured. The Ni peaks can be observed in the EDS spectrum for all the surface alloyed specimen and the spectrum for 200 μm Ni is shown in **Figure 8**.

Further, the results obtained by the EDAX analysis are reported in **Table 2**.

The Ni concentration values (wt %) reported in **Table 2** are plotted against the distance along the depth of the modified layer. **Figure 9** shows the Ni profiles for various coating thickness.

It can be observed from **Figure 9** that the Ni concentration is found to be the maximum on the surface of the modified layer for all the coating thickness. The Ni concentration decreases along the depth of the modified layer for all the coating thickness. It can be clearly observed that a gradient exists in the Ni concentration profile.

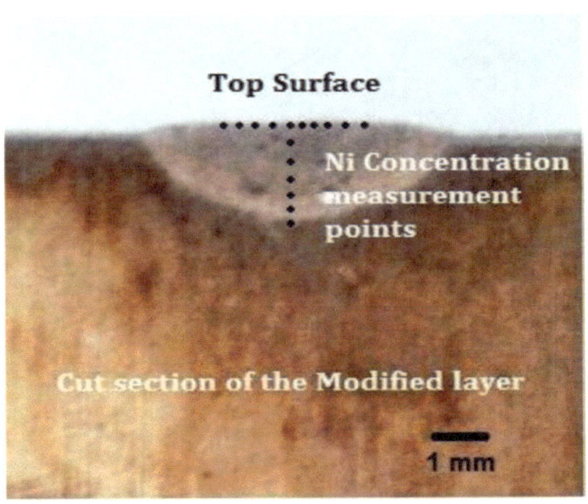


**Figure 7.**
*Ni concentration measurement points.*

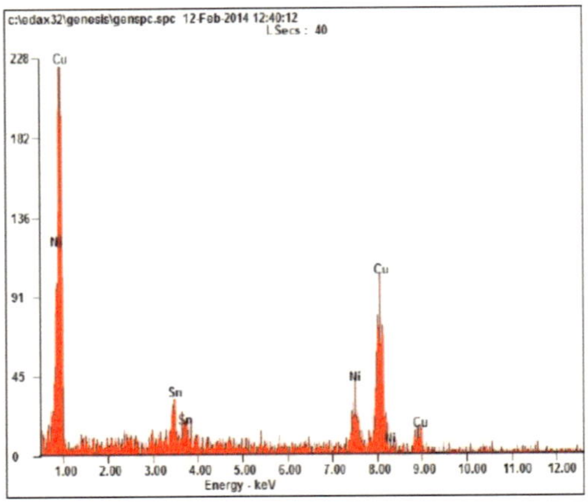


**Figure 8.**
*EDS spectrum for 200 μm Ni coated samples.*

| Depth from the top surface (mm) | Ni concentration for various coating thicknesses (wt %) | | | |
|---|---|---|---|---|
| | 80 μm | 120 μm | 160 μm | 200 μm |
| 0 | 5.03 | 8.53 | 13.61 | 17.81 |
| 0.3 | 4.9 | 8.39 | 12.92 | 14.88 |
| 0.6 | 4.48 | 7.15 | 10.18 | 13.65 |
| 0.9 | 3.26 | 6.46 | 7.86 | 10.41 |
| 1.2 | 2.95 | 5.83 | 6.85 | 7.35 |
| 1.5 | 2.15 | 4.36 | 5.02 | 5.89 |
| 1.8 | 0.87 | 2.48 | 3.15 | 4.26 |

**Table 2.**
*Ni concentration along the depth of the modified layer for four coating thickness.*

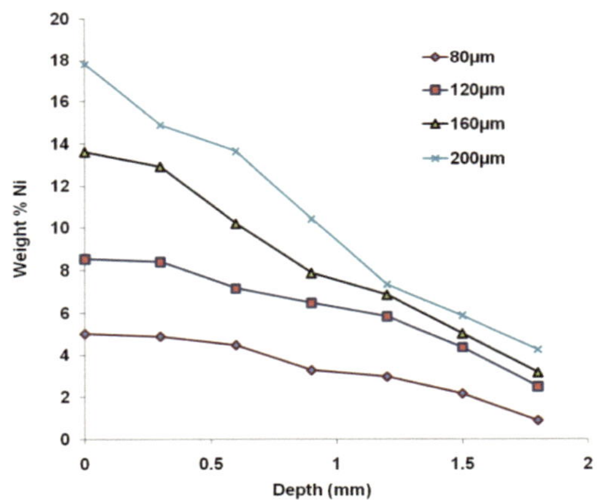


**Figure 9.**
*Ni concentration profile.*

## 4. Micro-hardness

The surface hardness values of the substrate and the surface alloyed specimens with varying Ni concentration were measured. Several readings were taken at different locations and an average value was calculated. The surface hardness increased from 120 HV for the substrate to 485 HV for the specimen surface alloyed with a Ni coating thickness of 200 μm. The average surface hardness values of the substrate and the Ni surface alloyed specimens are reported in **Table 3**.

The variation in the surface hardness with the Ni concentration is shown in **Figure 10**. It can be observed that the hardness increases with an increase in the Ni concentration. Ni contributes significantly to the hardness of the Cu-10Sn bronze alloy. The increase in the hardness is attributed to the presence of Ni in the solid solution. Hence, the hardening mechanism is solid solution strengthening.

Hardness values are measured at different points along the depth of the modified layer and are reported in **Table 4** and are represented graphically in **Figure 11**.

The hardness is found to decrease along the depth direction for all the surface alloyed specimen as shown in **Figure 11**. It can be concluded that a gradient exists in the hardness profile along the depth direction. The gradient so observed is attributed to the variation in the Ni concentration along the depth of the modified layer (refer to **Table 2**). The hardness is found to be the maximum for a concentration of 17.8 wt % Ni. It can be inferred that the hardness on the surface of the modified layer formed in the surface alloying process can be controlled by controlling the Ni concentration. Kac et al. [6] studied the structure and properties of Cu-10%Al-4%Fe-2%Mn bronze with an addition of Ti on the surface using laser as the heat source. They reported that a gradient exists in hardness along the depth direction of the modified layer. The observation obtained is consistent with that of Kac et al. [6].

**Figure 12** is a bar chart showing the hardness values obtained for the substrate, surface refined and the Ni surface alloyed specimens. It can be observed that the

| Alloy | Ni coating thickness (μm) | wt % Ni | Hardness (HV0.1) | |
|---|---|---|---|---|
| | | | Substrate | Surface alloyed with Ni |
| Cu-10Sn | 80 | 5.03 | 120 | 326 |
| Cu-10Sn | 120 | 8.53 | 120 | 379 |
| Cu-10Sn | 160 | 13.61 | 120 | 418 |
| Cu-10Sn | 200 | 17.81 | 120 | 485 |

**Table 3.**
*Hardness values.*

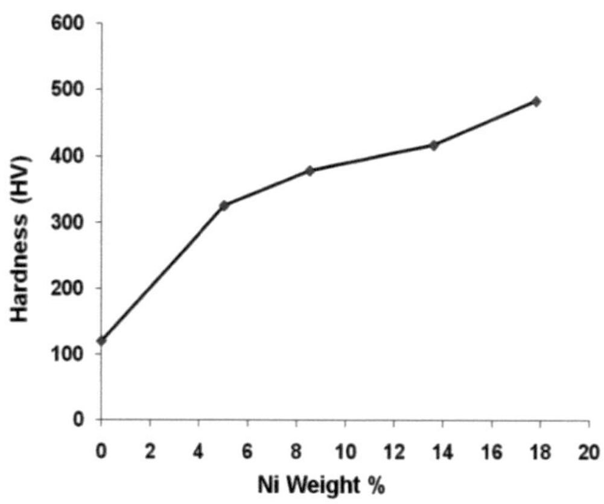


**Figure 10.**
*Hardness variation with Ni concentration.*

| Depth (mm) | Hardness for various Ni coating thickness (HV0.1) | | | |
|---|---|---|---|---|
| | 80 μm | 120 μm | 160 μm | 200 μm |
| 0 | 326 | 379 | 418 | 485 |
| 0.25 | 289 | 361 | 410 | 478 |
| 0.5 | 268 | 347 | 389 | 431 |
| 0.75 | 248 | 311 | 365 | 399 |
| 1 | 201 | 289 | 321 | 347 |
| 1.25 | 189 | 240 | 266 | 314 |
| 1.5 | 166 | 201 | 227 | 269 |
| 1.6 | 154 | 182 | 213 | 244 |
| 1.8 | 140 | 162 | 197 | 223 |
| 1.85 | 120 | 147 | 184 | 120 |
| 1.9 | 120 | 120 | 120 | 120 |

**Table 4.**
*Hardness along the depth of the modified layer for various Ni coating thickness.*

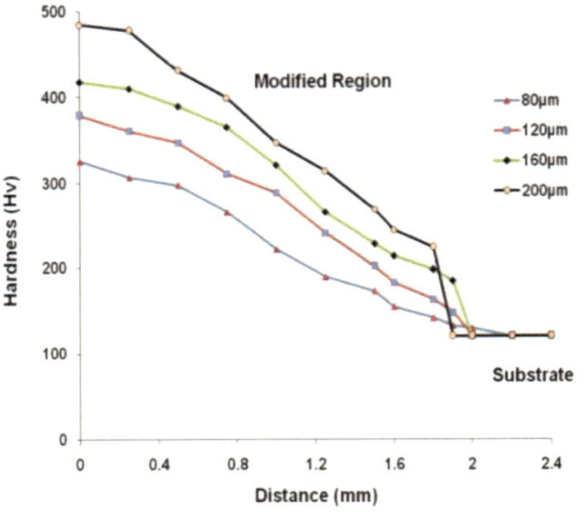


**Figure 11.**
*Hardness profile.*

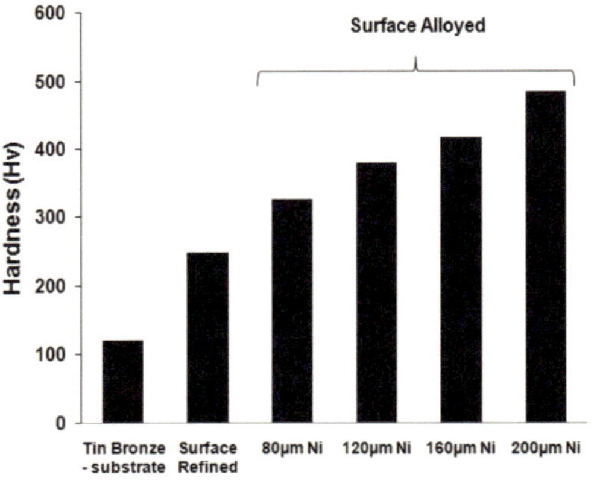


**Figure 12.**
*Hardness values—substrate, surface refined and Ni surface alloyed specimen.*

surface refining process and the surface alloying process significantly increases the hardness of the alloy. Increase in hardness observed in the surface refining process is attributed to the formation of fine grained microstructure due to rapid solidification in the surface refining process. However, the grain refinement occurs in the surface alloying process as shown in **Figure 6**. The addition of Ni in the surface alloying process also contributes to the improvement in the hardness of the alloyed specimen as shown in **Figure 11**. Hence, the increase in hardness is attributed to the grain refinement occurring in the surface alloying process and also to the Ni addition.

## 5. Wear behavior

A typical height loss vs. time plot for the Cu-10Sn modified alloy is shown in **Figure 13**. It can be observed that the height loss increases linearly with the sliding time. This behavior is in agreement with the results reported by Singh et al. [9] in the bulk alloys.

The wear results obtained for the substrate and the Ni surface alloyed samples are reported in **Table 5**.

It can be observed that the wear rate reduced significantly after surface alloying with Ni. The reduction in the wear rate is attributed to the increase in the hardness after Ni addition.

**Figure 14** is a bar chart showing the variation in the wear rate with the Ni concentration. It can be observed that the wear rate decreases with an increase in the Ni concentration. The minimum wear rate was obtained for the 17.8 wt % Ni. It can be concluded that the wear rate of the Cu-Sn bronze alloy can be reduced by surface alloying with Ni. The increased hardness due to the Ni addition is the reason behind the reduction in the wear rate.

**Figure 15** is a bar chart showing the wear rate obtained for the substrate, surface refined and the Ni surface alloyed specimens.

It can be observed from **Figure 15** that the surface refining process decreases the wear rate marginally and the surface alloying process remarkably decreases the wear rate of the Cu-10Sn bronze alloy. The reduction in the wear rate observed in the surface refining process is due to the increase in the hardness as a result of the grain refinement due to the faster cooling rate experienced. Further, it is to be noted that the refinement in the grain structure also occurs in the surface alloying process

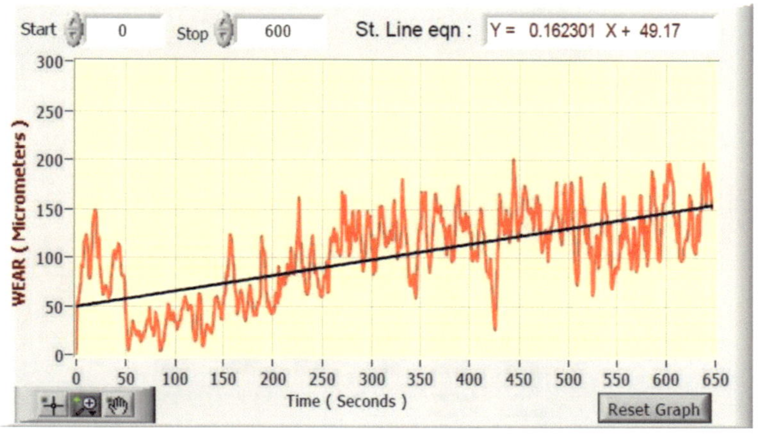


**Figure 13.**
*A typical wear plot.*

as discussed earlier. However, the Ni addition significantly contributes to the increase in the hardness that reduces the wear rate of the surface alloyed specimen. Hence, the reduction in the wear rate is attributed to both the grain refinement occurring in the surface alloying process and the Ni addition.

A typical image showing the wear tracks after the dry sliding test on pin-on-disc wear tester for the Ni surface alloyed specimen is shown in **Figure 16**.

It can be observed from **Figure 16** that the wear mechanism is of adhesive type. Zhang et al. [10] studied the dry sliding wear behavior in the bulk Cu-15Ni-8Sn alloy. They reported that the adhesive wear took place under the dry sliding test

| Substrate alloy | Ni coating thickness (μm) | Wear rate ($\times 10^{-4}$ $mm^3/m$) | |
|---|---|---|---|
| | | Substrate | Surface alloyed |
| Cu-10Sn | 80 | 18.40 | 13.70 |
| Cu-10Sn | 120 | 18.40 | 8.40 |
| Cu-10Sn | 160 | 18.40 | 4.5 |
| Cu-10Sn | 200 | 18.40 | 2.2 |

**Table 5.**
*Wear rate results.*

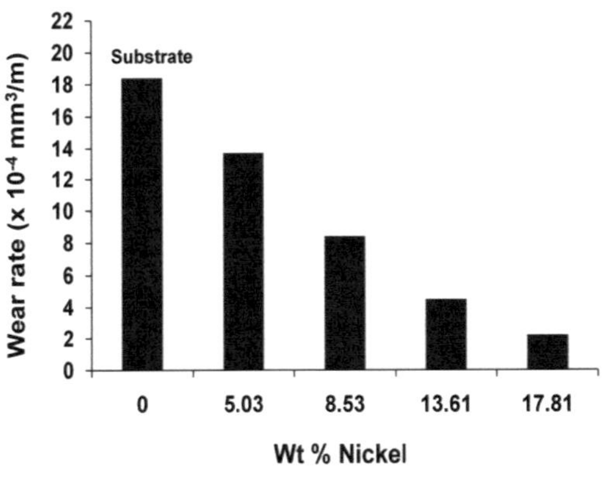


**Figure 14.**
*Variation in wear rate with Ni concentration.*

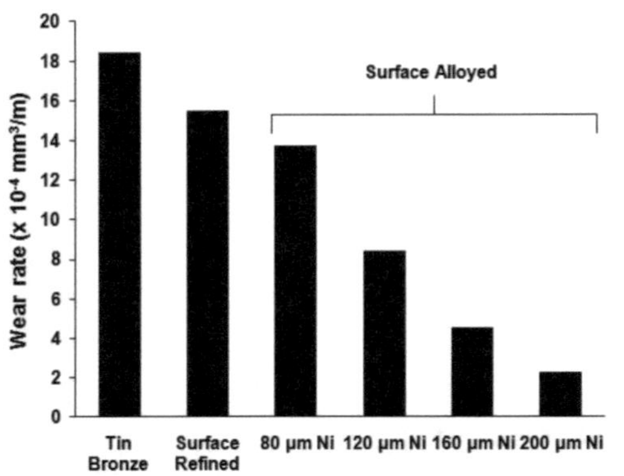


**Figure 15.**
*Wear rate—substrate, surface refined and surface alloyed specimens.*

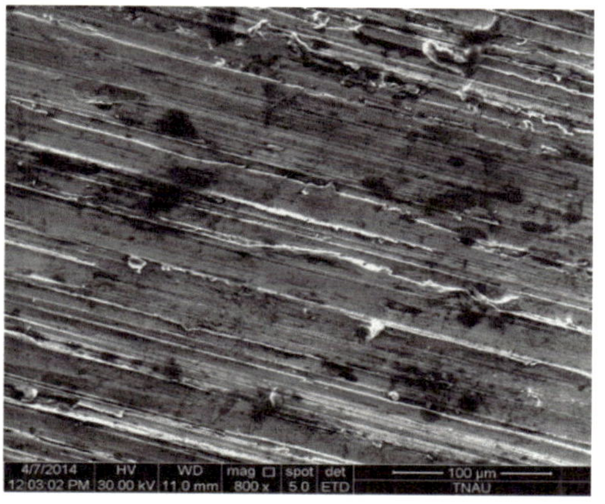


**Figure 16.**
*Wear tracks.*

conditions. It is to be noted that the compositions of the modified layer in this study are similar to the Cu-15Ni-8Sn alloy that was used in the study of Zhang et al. Therefore, the observation in the present study is consistent with that of Zhang et al. [11].

## 6. Coefficient of friction

Frictional force vs. time plot of Ni Surface Alloyed Cu-10Sn alloy is shown in **Figure 17**. The same trend was found for all the other Ni Surface alloyed specimens.

It can be noticed that the frictional force becomes constant after a short period of time and remains as such. The rapid increase in frictional force found initially is due to the uneven contact between the modified specimen and counter face material. The frictional force remains constant once perfect contact is achieved. A typical plot of coefficient of friction vs. time for the surface alloyed Cu-10Sn alloy is shown in **Figure 18**. The plot shows both transient period and single steady-state regime. The reasons for the transient behavior may be the effect of work-hardening and/or accumulation of debris as reported by Singh et al. [9].

The coefficient of friction obtained in this study for the substrate and the surface alloyed Cu-10Sn alloys are reported in **Table 6**. An average value of 0.23 was obtained as frictional coefficient after surface alloying process.

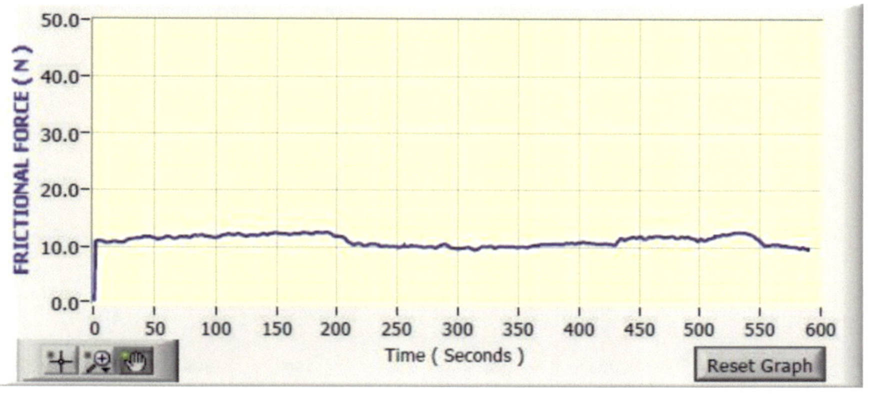


**Figure 17.**
*Frictional force vs. time plot.*

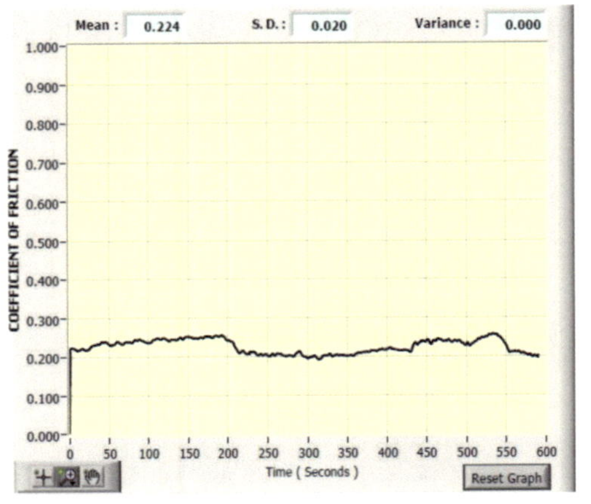


**Figure 18.**
*COF vs. time plot.*

| Substrate alloy | Ni coating thickness (μm) | Coefficient of friction | |
|---|---|---|---|
| | | Substrate | Surface alloyed |
| Cu-10Sn | 80 | 0.27 | 0.24 |
| Cu-10Sn | 120 | 0.27 | 0.22 |
| Cu-10Sn | 160 | 0.27 | 0.23 |
| Cu-10Sn | 200 | 0.27 | 0.22 |

**Table 6.**
*Frictional coefficients for substrate and surface alloyed specimens.*

| Substrate alloy | Ni coating thickness (μm) | Hardness (HV0.1) | Coefficient of friction |
|---|---|---|---|
| Cu-10Sn | 80 | 326 | 0.24 |
| Cu-10Sn | 120 | 379 | 0.22 |
| Cu-10Sn | 160 | 418 | 0.23 |
| Cu-10Sn | 200 | 485 | 0.22 |

**Table 7.**
*Coefficient of friction with hardness of the alloys.*

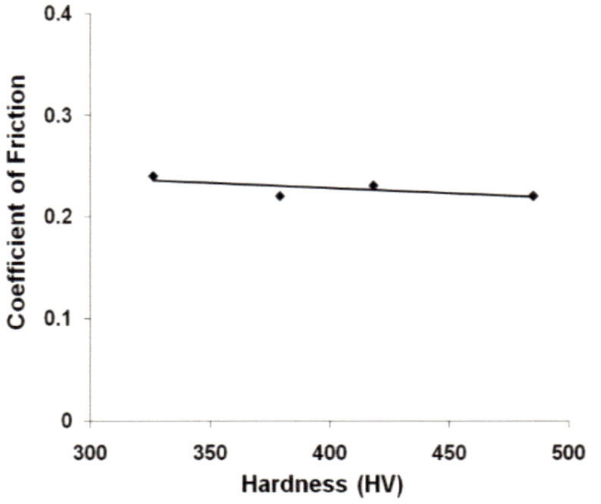


**Figure 19.**
*COF variation with hardness.*

The hardness values and the coefficient of friction of surface alloyed Cu-10Sn bronze alloys are reported in **Table 7**.

**Figure 19** shows the variation in coefficient of friction with hardness for the Cu-10Sn bronze alloys surface alloyed with Ni. It can be inferred that the COF remains almost a constant value irrespective of the hardness.

## 7. Conclusions

Cu-10Sn Bronze alloy was Surface Alloyed with Ni and a functionally graded Cu-Sn-Ni alloy with superior surface hardness and wear resistance was developed. Based on the results of this investigation, the following conclusions are drawn:

- Refinement in grain structure occurs in the surface alloying process.
- A gradient exists in the Ni concentration profile along the depth direction of the modified layer formed in the surface alloying process.
- Hardness can be significantly improved by surface alloying with Ni.
- A gradient exists in the hardness profile along the depth direction of the modified layer formed in the surface alloying process.
- Hardness on the surface can be easily controlled by controlling the Ni concentration on the surface of the modified layer.
- The wear rate was found to decrease with increase in hardness, a finding consistent with Archard's theory and that of the previous studies.
- Addition of Ni is found to be highly effective in increasing the hardness of the parent substrate and reducing the wear rate when compared to the addition of other alloying elements like Ti, TiC and Cr.
- The wear behavior is found to be adhesive in nature.
- The Coefficient of Friction of the Surface Alloyed specimen is found to remain constant irrespective of the hardness.

## Acknowledgements

The author is deeply indebted to Defence Research Development Organization, New Delhi (DRDO) for the financial support provided for conducting this project (Project No: ERIP/ER/1003934/M/01/1345 dated 26 July 2011).

## Author details

Cherian Paul[1*] and Ramasamy Sellamuthu[2]

1 Department of Mechanical Engineering, SAINTGITS College of Engineering, Kottayam, Kerala, India

2 Department of Mechanical Engineering, Amrita School of Engineering, Amrita Vishwa Vidyapeetham, Coimbatore, Tamilnadu, India

*Address all correspondence to: cherian.paul@saintgits.org

## References

[1] Benkisser G, Horn G, Semjonov S, Kaps R. Surface layer hardening of heterogeneous multialloy aluminium bronzes by laser beam melting. Metall. 1992;**46**(4):324-328

[2] Paul C, Sellamuthu R. An investigation on the effect of process parameters on microstructure, hardness and wear properties of surface modified Cu-Sn bronze alloy. Applied Mechanics and Materials. 2014;**592-594**:58-62

[3] Gadag SP, Galun R, Weisheit A, Mordike BL. Laser alloying of copper and its alloys. In: Laser Processing: Surface Treatment and Film Deposition. Vol. 307. 1996. pp. 359-377

[4] Yilbas BS, Matthews A, Leyland A, Karatas A, Akhtar SS, Abdul Aleem BJ. Laser surface modification treatment of aluminium bronze with B4C. Applied Surface Science. 2012;**263**:804-809

[5] Yilbas BS, Akhtar SS, Karatas C. Laser notriding of the surface of phosphor bronze. International Journal of Advanced Manufacturing Technology. 2013;**65**:1553-1565

[6] Kac S, Kusinski J. Structure and properties of the bronze laser alloyed with titanium. Applied Surface Science. 2007;**253**:7895-7898

[7] Viswanadham CS, Goswami GL, Galun R, Mordike BL. Laser melt injection of TiC particles into aluminium bronze. Laser in engineering. 2006;**16**:207-213

[8] Majumdar JD, Manna I. Laser processing of materials. Sadhana. 2003;**28**(3&4):495-562

[9] Singh JB, Cai W, Bellon P. Dry sliding of Cu-15 wt%Ni-8 wt%Sn bronze: Wear behaviour and microstructures. Wear. 2007;**263**:830-841

[10] Zhang SZ, Jiang B, Ding WJ. Wear of Cu-15Ni-8Sn spinodal alloy. Wear. 2009;**264**(3-4):199-203

[11] Zhang SZ, Jiang B, Ding WJ. Dry sliding wear of Cu-15Ni-8Sn alloy. Tribology International. 2010;**43**(1-2):64-68

Chapter 6

# Axisymmetric Indentation Response of Functionally Graded Material Coating

*Tie-Jun Liu*

## Abstract

In this chapter, the indentation response of the functionally graded material (FGM) coating is considered due to the contact between the coating and axisymmetric indenter. The mechanical properties of FGM coating is assumed to vary along the thickness direction. Three kinds of models are applied to simulate the variation of elastic parameter in the FGM coating based on the cylindrical coordinate system. The axisymmetric frictionless and partial slip contact problems are reduced to a set of Cauchy singular integral equations that can be numerically calculated by using the Hankel integral transform technique and the transfer matrix method. The effect of gradient of coating on the distribution of contact stress is presented. The present investigation will provide the guidance for the indentation experiment of coating.

**Keywords:** functionally graded material, coating, indentation, axisymmetric contact

## 1. Introduction

Functionally graded material (FGM) [1] which is new kind of nonhomogeneous composite material has many predominant properties, so it has been widely used in many fields. In recent years, many researchers have conducted the experiment to prove that FGM used as coatings can resist the contact deformation and reduce the interface damage [2], so it is very important to study the indentation response of FGM coating. Because FGM are composites whose material properties vary gradually along a coordinate axis, the governing equations which represent the mechanical behaviors of the materials are very difficult to solve. Researchers usually describe the properties of FGM according to some specific functional forms such as exponential functions and power law functions of elastic modulus [3, 4]. By assuming the elastic modulus of FGM varying as exponential function form, Guler and Erdogan [5, 6] studied the two-dimensional contact problem of functionally graded coatings. Liu et al. [7, 8] investigated the axisymmetric contact problem of FGM coating and interfacial layer with exponentially varying modulus by using the singular integral equation. The axisymmetric problems for a nonhomogeneous elastic layer in which the shear modulus follows the power law function are taken into account by Jeon et al. [9]. Because solving the controlling equations of FGM is difficult, the contact problem of FGM is limited to assume the elastic modulus varying as some specific functional forms. To eliminate this disadvantage, Ke and Wang [10, 11] applied the linear multilayered (LML) model to simulate the FGM with arbitrarily varying elastic parameter. Based on the model, some two-dimensional contact problems are studied. The axisymmetric contact problem of FGM coating with arbitrary spatial variation of material properties is considered by making use of the extended linear multilayered model [12, 13]. Recently, a piece-wise exponential multilayered (PWEML) model [14] is presented to solve the frictionless contact problem of FGM with the shear modulus of the coating varying in the power law form. Subsequently, Liu and Li [15] applied the model to solve the two-dimensional adhesive contact problem.

When two bodies are brought together under the applied force, contact occurs at interface. Hertz [16] first considers the frictionless contact problem between elastic bodies. Researchers obtained the classical solution to the indentation problem under the flat, cylindrical, and cone punch based on Hertz's theory [17]. The contact tractions and displacement field can be given to characterize the mechanical properties of various materials. Liu et al. [7, 12, 14] solved the axisymmetric frictionless contact problem for FGM coating by using the singular integral equation. They discussed the effect of the gradient of FGM coating on the indentation response. Because the materials of the two contact solids are dissimilar, the slip will take place at the contact surface. If slip is opposed by friction, the contact region is divided into two parts: the stick region and the slip region. Spence [18] gives the contact stress fields in homogeneous materials by assuming a self-similarity at each stage of finite friction contact when the normal load monotonically increases. Ke and Wang [19] solved the two-dimensional contact problem with finite friction for FGM coating. Liu et al. [13] considered the axisymmetric partial slip contact problem of a graded coating. When the coefficient of friction is sufficiently large, slip might be prevented entirely. The self-similar solution to nonslip contact problems with incremental loading was considered by Spence [20]. Goodman [21] investigated the axisymmetric contact problem with full stick when elastically dissimilar spheres are pressed together. Mossakovski [22] studied contact with adhesion for the elastic bodies under condition of adhesion. Norwell et al. [23] adopt an iteration method to solve the coupled equations which can describe the partial slip contact problem.

In this chapter, the axisymmetric frictionless and partial slip contact problems for FGM coating are considered. The basic formulation for nonhomogeneous material layer with elastic parameter varying along the thickness direction is given in Section 2. Based on the basic formulations for nonhomogeneous layer, three types of computational model for FGM coating are introduced in Section 3 for axisymmetric contact problem. The displacement and stress components in the transform domain are gained by using the Hankel transform technology and transfer matrix method. In Section 4, we will investigate the solution for the axisymmetric frictionless and partial slip contact problems. The indentation response of FGM coating under frictionless and frictional condition will be discussed in Section 5. Finally, we will depict some conclusions on the axisymmetric indentation response of FGM coating.

## 2. Basic formulations for nonhomogeneous material layer

For the present axisymmetric problem, the strain components, stress-strain relations, and the equilibrium equations in the radial and axial directions disregarding the body forces are given by the following relations [7]:

$$\varepsilon_{rr} = \frac{\partial u}{\partial r}, \tag{1a}$$

$$\varepsilon_{\theta\theta} = \frac{u}{r}, \tag{1b}$$

$$\varepsilon_{zz} = \frac{\partial w}{\partial z}, \tag{1c}$$

$$2\varepsilon_{rz} = \frac{\partial u}{\partial z} + \frac{\partial w}{\partial r}. \tag{1d}$$

$$\sigma_{rr} = (\lambda(z) + 2\mu(z))\frac{\partial u}{\partial r} + \lambda(z)\left(\frac{u}{r} + \frac{\partial w}{\partial z}\right), \tag{2a}$$

$$\sigma_{\theta\theta} = (\lambda(z) + 2\mu(z))\frac{u}{r} + \lambda(z)\left(\frac{\partial u}{\partial r} + \frac{\partial w}{\partial z}\right), \tag{2b}$$

$$\sigma_{zz} = (\lambda(z) + 2\mu(z))\frac{\partial w}{\partial z} + \lambda(z)\left(\frac{\partial u}{\partial r} + \frac{u}{r}\right), \tag{2c}$$

$$\sigma_{rz} = \mu(z)\left(\frac{\partial u}{\partial z} + \frac{\partial w}{r}\right). \tag{2d}$$

$$\frac{\partial \sigma_{rr}}{\partial r} + \frac{\partial \sigma_{rz}}{\partial z} + \frac{1}{r}(\sigma_{rr} - \sigma_{\theta\theta}) = 0, \tag{3a}$$

$$\frac{\partial \sigma_{rz}}{\partial r} + \frac{\partial \sigma_{zz}}{\partial z} + \frac{1}{r}\sigma_{rz} = 0. \tag{3b}$$

in which $r$ and $z$ are the variables of the cylindrical coordinate system; $\varepsilon_{rr}$, $\varepsilon_{\theta\theta}$, $\varepsilon_{zz}$, and $\varepsilon_{rz}$ are the strain components; $u$ and $w$ are the displacement components in the radial and axial directions; $\sigma_{rr}$, $\sigma_{\theta\theta}$, $\sigma_{zz}$, and $\sigma_{rz}$ are the stress components; $\lambda(z)$ and $\mu(z)$ are Lame's constants which vary along the $z$-axis direction.

## 3. Computational models for FGM coating

The properties of nonnonhomogeneous material may vary arbitrarily along a certain spatial direction, which makes the solution of contact problem very difficult in mathematics. In the present work, we adopt three methods to model the axisymmetric FGM layer based on the cylindrical coordinate system. First, exponential function (EF) model [7] is used to assume the elastic modulus of the FGM layer that varies as the exponential function. Second, the linear multi-layered (LML) model [12] is applied to simulate the FGM layer with arbitrarily varying material modulus, and Poisson's ratio is chosen as 1/3. The model divided FGM layer into a series of sub-layers in which the shear modulus varies as linear function form. The shear modulus is taken to be continuous at the sub-interfaces and equal to their real values. Third, the piecewise exponential multilayered (PWEML) model [14] is employed in modeling the functionally graded material layer with arbitrary spatial variation of material properties. In this model, the functionally graded layer is cut into several sub-layers where the elastic parameter varies according to the exponential function form. Three types of computational model for FGM coating are the following.

### 3.1 Exponential function model

In **Figure 1(a)**, the shear modulus of the functionally graded coating can be described by

$$\mu(z) = \mu_0 e^{\alpha z} \tag{4}$$

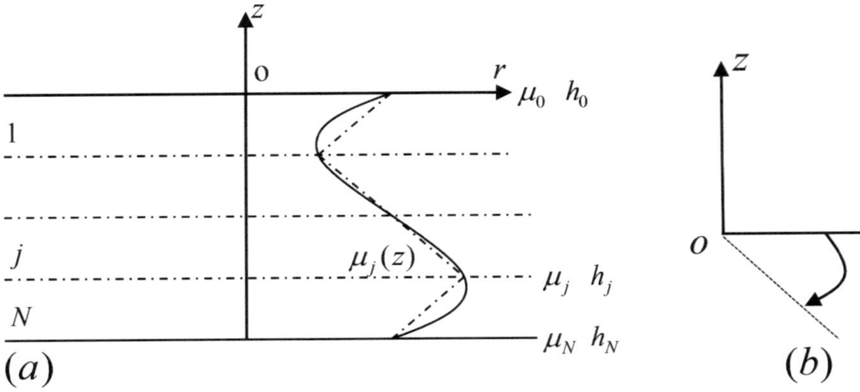


**Figure 1.**
*The linear mutli-layered model for the functionally graded coating (a) and the cylindrical coordinate system (b).*

where $\alpha = h_0^{-1} \log(\mu_0/\mu*)$ is a constant characterizing the material inhomogeneity with $\mu_0$ being the value of $\mu(z)$ at the surface, i.e., $\mu_0 = \mu(h_0)$. $\mu_0$ and $\mu*$ are related by

$$\mu_0 = \mu * e^{\alpha h_0} \tag{5}$$

Substituting Eqs. (2) and (5) into Eq. (3), we obtain

$$(k+1)\left\{\frac{\partial^2 u}{\partial r^2} + \frac{1}{r}\frac{\partial u}{\partial r} - \frac{1}{r^2}u + \frac{\partial^2 w}{\partial r \partial z}\right\} + (k-1)\alpha\left(\frac{\partial u}{\partial z} + \frac{\partial w}{\partial r}\right) + (k-1)\left\{\frac{\partial^2 u}{\partial z^2} - \frac{\partial^2 w}{\partial r \partial z}\right\} = 0, \tag{6a}$$

$$(k+1)\left\{\frac{\partial^2 u}{\partial r \partial z} + \frac{1}{r}\frac{\partial u}{\partial z} + \frac{\partial^2 w}{\partial z^2}\right\} - (k-1)\left(\frac{\partial^2 u}{\partial r \partial z} - \frac{\partial^2 w}{\partial r^2}\right) - \frac{(k-1)}{r}\left\{\frac{\partial u}{\partial z} - \frac{\partial w}{\partial r}\right\} + (3-k)\alpha\left(\frac{\partial u}{\partial r} + \frac{u}{r}\right) + (k+1)\alpha\frac{\partial w}{\partial z} = 0 \tag{6b}$$

where $k = 3 - 4\nu$ and $\nu$ is Poisson's ratio.

In order to solve Eq. (6), we use the technique of Hankel integral transform. The Hankel transform and its inversion are defined as

$$\langle \tilde{\omega}(s,z)\rangle_p = \int_0^\infty \omega(s,z) r J_p(sr) dr, \tag{7a}$$

$$\omega_p(r,z) = \int_0^\infty \langle \tilde{\omega}(s,z)\rangle_p s J_p(sr) ds \tag{7b}$$

where the bar $\sim$ indicates Hankel transform; $\langle\rangle_p$ is the $p$th-order Hankel transform; and $J_p$ is the $p$th-order Bessel function of the first kind.

By using the Hankel transform and defining $D = d/dz$, Eq. (6) can be expressed as

$$\left\{(k-1)D^2 + \alpha(k-1)D - (k+1)s^2\right\}\langle\tilde{u}\rangle_1 - \{2sD + \alpha s(k-1)\}\langle\tilde{w}\rangle_0 = 0, \tag{8a}$$

$$\{2sD + \alpha(3-k)s\}\langle\tilde{u}\rangle_1 + \left\{(k+1)D^2 + \alpha(k+1)D - (k-1)s^2\right\}\langle\tilde{w}\rangle_0 = 0, \quad (8b)$$

The solution of the differential Eqs. (8) is given by [7]

$$\langle\tilde{u}\rangle_1 = A_3(s)e^{m_1 z} + A_4(s)e^{m_2 z} + A_5(s)e^{m_3 z} + A_6(s)e^{m_4 z} \quad (9a)$$

$$\langle\tilde{w}\rangle_0 = A_3(s)a_1e^{m_1 z} + A_4(s)a_2e^{m_2 z} + A_5(s)a_3e^{m_3 z} + A_6(s)a_4e^{m_4 z} \quad (9b)$$

where

$$a_i = -\frac{2sm_i + s\alpha(3-k)}{(k+1){m_i}^2 + \alpha(k+1)m_i - (k-1)s^2}, (i = 1, \ldots.., 4)$$

and

$$m_1 = -\frac{\alpha}{2} + \frac{1}{2}\left\{\alpha^2 + 4s^2 + i4\sqrt{\frac{3-k}{k+1}}\alpha s\right\}^{1/2},$$

$$m_2 = -\frac{\alpha}{2} - \frac{1}{2}\left\{\alpha^2 + 4s^2 + i4\sqrt{\frac{3-k}{k+1}}\alpha s\right\}^{1/2},$$

$$m_3 = -\frac{\alpha}{2} + \frac{1}{2}\left\{\alpha^2 + 4s^2 - i4\sqrt{\frac{3-k}{k+1}}\alpha s\right\}^{1/2},$$

$$m_4 = -\frac{\alpha}{2} - \frac{1}{2}\left\{\alpha^2 + 4s^2 - i4\sqrt{\frac{3-k}{k+1}}\alpha s\right\}^{1/2}.$$

According to Hooke's law and strain-displacement relations, stress components may be expressed as

$$\frac{\kappa-1}{\mu(z)}\langle\tilde{\sigma}_{zz}\rangle_0 = \sum_{i=1}^{4}\{(\kappa+1)m_i a_i + (3-k)s\}A_{i+2}e^{m_i z} \quad (10a)$$

$$\frac{1}{\mu(z)}\langle\tilde{\sigma}_{rz}\rangle_1 = \sum_{i=1}^{4}\{m_i - a_i s\}A_{i+2}e^{m_i z} \quad (10b)$$

For a homogeneous layer without the gradient, the gradient index $\alpha$ in Eqs. (9) and (10) equals to 0.

### 3.2 Linear multi-layered model

Consider the linear multi-layered model shown in **Figure 1**. The shear modulus of the functionally graded coating can be described by an arbitrary continuous function of $z$, $\mu(z)$, with boundary values $\mu(h_0) = \mu_0$. Poisson's ratio $v$ is taken as 1/3. The linear multilayered model divides functionally graded coating into $N$ sub-layers. The shear modulus $\mu(z)$ in each sub-layer is assumed to take the following form:

$$\mu(z) \approx \mu_j(z) = c_j(1 + z/b_j) = c_j\left(\frac{z*}{b_j}\right), h_j \le z \le h_{j-1}, j = 1, 2 \ldots N \quad (11)$$

where $z* = z + b$ and $\mu_j$ is equal to the real value of the shear modulus at the sub-interfaces, $z = h_j$, i.e., $\mu_j(h_j) = \mu(h_j)$, which lead to

$$b_j = \frac{\mu_{j-1}h_j - \mu_j h_{j-1}}{\mu_j - \mu_{j-1}}, \tag{12a}$$

$$c_j = \frac{\mu_j}{1 + h_j/b_j}. \tag{12b}$$

As in [12], introduces two potential functions to write the displacement components $u_j$ and $w_j$ in each sub-layer:

$$u_j = (z + b_j)\partial f_j/\partial r + \partial\phi_j/\partial r \tag{13a}$$

$$w_j = (z + b_j)\partial f_j/\partial z - f_j + \partial\phi_j/\partial z, \ (h_j \le z \le h_{j-1}). \tag{13b}$$

By making use of Eqs. (1), (2), and (13), the equilibrium equations (3) are represented as [12].

$$\nabla^2\phi_j + \frac{1}{z*}\partial\phi_j/\partial z* = 0, \tag{14a}$$

$$\nabla^2 f_j + \frac{1}{z*}\partial f_j/\partial z* = \frac{1}{2z*^2}\partial\phi_j/\partial z. \tag{14b}$$

where

$$\nabla^2 = \frac{1}{r}\partial(r\partial/\partial r)/\partial r + \partial^2/\partial z*^2.$$

Then the displacement and stress components given by Eqs. (13) and (2) are given by

$$u_j = z*\partial f_j/\partial r + \partial\phi_j/\partial r, \ (h_j \le z \le h_{j-1}) \tag{15a}$$

$$w_j = z*\partial f_j/\partial z* - f_j + \partial\phi_j/\partial z*, \ (h_j \le z \le h_{j-1}) \tag{15b}$$

$$\sigma_{rrj} = 2c_j\left(\frac{z*}{b_j}\right)\left(z*\partial^2 f_j/\partial r^2 + \partial^2\phi_j/\partial r^2 - \partial f_j/\partial z* - \frac{1}{2z*}\partial\phi_j/\partial z*\right), \tag{16a}$$

$$\sigma_{\theta\theta j} = 2c_j\left(\frac{z*}{b_j}\right)\left(z*\frac{1}{r}\partial f_j/\partial r + \frac{1}{r}\partial\phi_j/\partial r - \partial f_j/\partial z* - \frac{1}{2z*}\partial\phi_j/\partial z*\right), \tag{16b}$$

$$\sigma_{zzj} = 2c_j\left(\frac{z*}{b_j}\right)\left(z*\partial^2 f_j/\partial r^2 + \partial^2\phi_j/\partial z*^2 - \partial f_j/\partial z* - \frac{1}{2z*}\partial\phi_j/\partial z*\right), \tag{16c}$$

$$\sigma_{rzj} = 2c_j\left(\frac{z*}{b_j}\right)\left(z*\partial^2 f_j/\partial r\partial z* + \partial^2\phi_j/\partial r\partial z*\right). \tag{16d}$$

Applying Hankel transformation Eqs. (7a)–(14), we obtain the solutions for displacement functions $\phi_j$ and $f_j$ in each sub-layer:

$$\langle\tilde{\phi}_j\rangle_0 = A_{j1}(s)I_0(sz*) + A_{j2}(s)K_0(sz*), \tag{17a}$$

$$\langle\tilde{f}_j\rangle_0 = A_{j3}(s)I_0(sz*) + A_{j4}(s)K_0(sz*) + \frac{s}{2}\{A_{j1}(s)I_1(sz*) - A_{j2}(s)K_1(sz*)\} \tag{17b}$$

where $I_0$, $I_1$, $K_0$, and $K_1$ are modified Bessel functions of the 0th and 1th order.

Applying Hankel transform to Eqs. (15) and (16), we get

$$\begin{aligned}\langle \tilde{u}_{rj} \rangle_1 &= -sz * \langle \tilde{f}_j \rangle_0 - s\langle \tilde{\phi}_j \rangle_0 \\ &= \left\{ -\frac{s^2}{2} z * I_1(sz*) - sI_0(sz*) \right\} A_{j1}(s) + \left\{ \frac{s^2}{2} z * K_1(sz*) - sK_0(sz*) \right\} A_{j2}(s) \\ &\quad -sz * I_0(sz*) A_{j3}(s) - sz * K_0(sz*) A_{j4}(s)\end{aligned} \tag{18a}$$

$$\begin{aligned}\langle \tilde{w}_j \rangle_0 &= z * d\langle \tilde{f}_j \rangle_0 / dz* - \langle \tilde{f}_j \rangle_0 + d\langle \tilde{\phi}_j \rangle_0 / dz* \\ &= \frac{s^2}{2} z * I_0(sz*) A_{j1}(s) + \frac{s^2}{2} z * K_0(sz*) A_{j2}(s) + \{ z * sI_1(sz*) - I_0(sz*) \} A_{j3}(s) \\ &\quad + \{ -z * sK_1(sz*) - K_0(sz*) \} A_{j4}(s)\end{aligned} \tag{18b}$$

$$\begin{aligned}\langle \tilde{\sigma}_{zzj} \rangle_0 = 2c_j \left( \frac{z*}{b_j} \right) &\left\{ \frac{s^3}{2} z * I_1(sz*) A_{j1}(s) - \frac{s^3}{2} z * K_1(sz*) A_{j2}(s) \right. \\ &\left. + \left( z * s^2 I_0(sz*) - 2sI_1(sz*) \right) A_{j3}(s) + \left( z * s^2 K_0(sz*) + 2sK_1(sz*) \right) A_{j4}(s) \right\}\end{aligned} \tag{18c}$$

$$\begin{aligned}\langle \tilde{\sigma}_{rzj} \rangle_1 &= 2c_j \left( \frac{z*}{b_j} \right) \left\{ \left( -\frac{s^3}{2} z * I_0(sz*) - \frac{s^2}{2} I_1(sz*) \right) A_{j1}(s) \right. \\ &\left. + \left( -\frac{s^3}{2} z * K_0(sz*) + \frac{s^2}{2} K_1(sz*) \right) A_{j2}(s) - s^2 z * I_1(sz*) A_{j3}(s) + s^2 z * K_1(sz*) A_{j4}(s) \right\}\end{aligned} \tag{18d}$$

### 3.3 Piece wise exponential multi-layered model

Piece wise exponential multi-layered model divides functionally graded coatings into $N$ sub-layers as shown in **Figure 2**. The shear modulus $\mu(z)$ in each sub-layer is assumed to vary as an exponential function form:

$$\mu(z) \approx \mu_j(z) = a_j e^{b_j z}, h_j \le z \le h_{j-1}, j = 1, 2, \ldots\ldots N \tag{19a}$$

$$\mu(h_j) = \mu_j(h_j), \tag{19b}$$

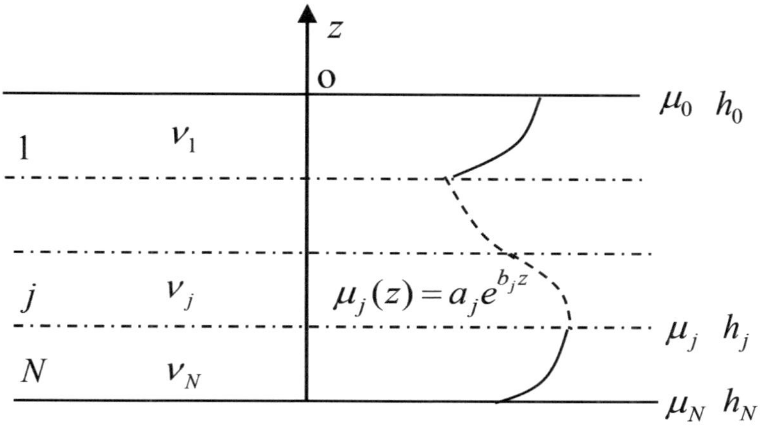


**Figure 2.**
*Piece wise exponential multi-layered model for the graded coating.*

in which:

$$a_j = \mu(h_j)e^{-\ln[\mu(h_{j+1})/\mu(h_j)]h_j/(h_{j+1}-h_j)}, b_j = \ln[\mu(h_{j+1})/\mu(h_j)]/(h_{j+1}-h_j)$$

and $h_j$ is the $z$ coordinate at the end of layer $j$. Poisson's ratio in each sub-layer is assumed to be a constant $v_j$.

In each sub-layer ($j = 1, 2, \ldots\ldots N$), the equilibrium equations are represented as [14]

$$(k_j+1)\left\{\frac{\partial^2 u_j}{\partial r^2}+\frac{1}{r}\frac{\partial u_j}{\partial r}-\frac{1}{r^2}u_j+\frac{\partial^2 w_j}{\partial r\partial z}\right\}+(k_j-1)b_j\left(\frac{\partial u_j}{\partial z}+\frac{\partial w_j}{\partial r}\right)$$
$$+(k_j-1)\left\{\frac{\partial^2 u_j}{\partial z^2}-\frac{\partial^2 w_j}{\partial r\partial z}\right\}=0, \tag{20a}$$

$$(k_j+1)\left\{\frac{\partial^2 u_j}{\partial r\partial z}+\frac{1}{r}\frac{\partial u_j}{\partial z}+\frac{\partial^2 w_j}{\partial z^2}\right\}-(k_j-1)b_j\left(\frac{\partial^2 u_j}{\partial r\partial z}-\frac{\partial^2 w_j}{\partial r^2}\right)-\frac{(k_j-1)}{r}\left\{\frac{\partial u_j}{\partial z}-\frac{\partial w_j}{\partial r}\right\}$$
$$+(3-k)b_j\left(\frac{\partial u_j}{\partial r}+\frac{u_j}{r}\right)+(k_j+1)b_j\frac{\partial w_j}{\partial z}=0 \tag{20b}$$

where $u_j$ and $w_j$ are the displacement components in the radial and $z$ axial directions in layer $j$ and $k_j = 3 - 4v_j$.

The solution of differential equations (20) may be expressed as [7]

$$\langle \tilde{u}_j(s,z)\rangle_1 = A_{j1}(s)e^{m_{j1}z}+A_{j2}(s)e^{m_{j2}z}+A_{j3}(s)e^{m_{j3}z}+A_{j4}(s)e^{m_{j4}z} \tag{21a}$$

$$\langle \tilde{w}_j(s,z)\rangle_0 = A_{j1}(s)c_{j1}e^{m_{j1}z}+A_{j2}(s)c_{j2}e^{m_{j2}z}+A_{j3}(s)c_{j3}e^{m_{j3}z}+A_{j4}(s)c_{j4}e^{m_{j4}z} \tag{21b}$$

where $A_{j1}$–$A_{j4}$ are unknown constants to be solved in layer $j$.

$$c_{ji} = -\frac{2sm_{ji}+sb_j(3-k_j)}{(k_j+1)m_{ji}{}^2+b_j(k+1)m_{ji}-(k_j-1)s^2}, (i=1,2,3,4)$$

$$m_{j1} = -\frac{b_j}{2}+\frac{1}{2}\left\{b_j{}^2+4s^2+i4\sqrt{\frac{3-k_j}{k_j+1}}b_js\right\}^{1/2},$$

$$m_{j2} = -\frac{b_j}{2}-\frac{1}{2}\left\{b_j{}^2+4s^2+i4\sqrt{\frac{3-k_j}{k_j+1}}b_js\right\}^{1/2},$$

$$m_{j3} = -\frac{b_j}{2}+\frac{1}{2}\left\{b_j^2+4s^2-i4\sqrt{\frac{3-k_j}{k_j+1}}b_js\right\}^{1/2},$$

$$m_{j4} = -\frac{b_j}{2}-\frac{1}{2}\left\{b_j^2+4s^2-i4\sqrt{\frac{3-k_j}{k_j+1}}b_js\right\}^{1/2}.$$

According to Hooke's law and strain-displacement relations, stress components may be expressed as

$$\frac{k_j-1}{\mu_j(z)}\langle\tilde{\sigma}_{zzj}(s,z)\rangle_0 = \sum_{i=1}^{4}\{(k_j+1)m_{ji}c_{ji}+(3-k)s\}A_{ji}e^{m_{ji}z} \tag{22a}$$

$$\frac{1}{\mu_j(z)}\langle\tilde{\sigma}_{rzj}(s,z)\rangle_1 = \sum_{i=1}^{4}\{m_{ji} - c_{ji}s\}A_{ji}e^{m_{ji}z} \tag{22b}$$

## 4. Solution for the axisymmetric frictionless and partial slip contact problem

In this section, we will solve axisymmetric contact and fretting problem for the functionally graded coating bonded to the homogeneous half-space under the spherical indenter. A functionally graded coated half-space subjected to normal and radical distributed external loads is shown in **Figure 3**. The stresses and displacements are continuous at the interfaces, $z = 0$, which state.

$$u_2(r,0) - u_1(r,0) = 0, \tag{23a}$$

$$w_2(r,0) - w_1(r,0) = 0, \tag{23b}$$

$$\sigma_{2zz}(r,0) - \sigma_{1zz}(r,0) = 0, \tag{23c}$$

$$\sigma_{2rz}(r,0) - \sigma_{1rz}(r,0) = 0. \tag{23d}$$

And along the coating surface, $z = h_0$, we have

$$\sigma_{1zz}(r,h_0) = p(r)\ (0 \le r \le a), \tag{24a}$$

$$\sigma_{1zz}(r,h_0) = 0\quad (a < r < \infty), \tag{24b}$$

$$\sigma_{1rz}(r,h_0) = q(r)\ (0 \le r \le a), \tag{24c}$$

$$\sigma_{1rz}(r,h_0) = 0\quad (a < r < \infty) \tag{24d}$$

in which $i = 1$ refers to the graded coating and $i = 2$ refers to the homogeneous half-space. $p(r)$ and $q(r)$ are normal contact tractions and shear stress, respectively.

By using the Hankel integral transform technique and transfer matrix method, the surface displacement components can be expressed as

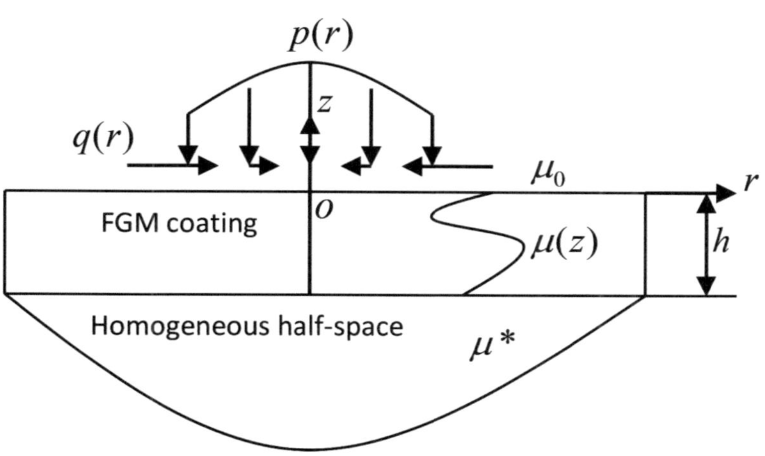


**Figure 3.**
*A functionally graded coated half-space subjected to normal and radical distributed external loads.*

$$w_0(r) = \int_0^a p(t)t \int_0^\infty sM_{11}(s,h_0)J_0(st)J_0(sr)dsdt + \int_0^a q(t)t \int_0^\infty sM_{12}(s,h_0)J_1(st)J_0(sr)dsdt \tag{25a}$$

$$u_0(r) = \int_0^a p(t)t \int_0^\infty sM_{21}(s,h_0)J_0(st)J_1(sr)dsdt + \int_0^a q(t)t \int_0^\infty sM_{22}(s,h_0)J_1(st)J_1(sr)dsdt \tag{25b}$$

where $J_0(.)$ and $J_1(.)$ are Bessel functions and

$$\begin{bmatrix} M_{11}(s,h_0) & M_{12}(s,h_0) \\ M_{21}(s,h_0) & M_{22}(s,h_0) \end{bmatrix} = \frac{1}{2\mu_0}\mathbf{B_3M},$$

$$\mathbf{M} = [T_1(h_0+b_1)][\overline{V}_1]\left\{[B][T_1(h_0+b_1)][\overline{V}_1]\right\}^{-1},$$

$$\mathbf{B_3} = \begin{bmatrix} 0 & 1 & 0 & 0 \\ 1 & 0 & 0 & 0 \end{bmatrix}, \mathbf{C} = \begin{bmatrix} 1 & 0 \\ 0 & 1 \end{bmatrix}.$$

where $M_{ij}(s,h_0)$ is the kernel function (see Ref. [13]).

Considering the asymptotic behavior of Bessel functions for large arguments [13], one may prove

$$\lim_{s\to\infty} \begin{bmatrix} sM_{11}(s,h_0) & sM_{12}(s,h_0) \\ sM_{21}(s,h_0) & sM_{22}(s,h_0) \end{bmatrix} = \begin{bmatrix} \alpha_{11} & \alpha_{12} \\ \alpha_{21} & \alpha_{22} \end{bmatrix} = \begin{bmatrix} \dfrac{1-v}{\mu_0} & \dfrac{1-2v}{2\mu_0} \\ \dfrac{1-2v}{2\mu_0} & \dfrac{1-v}{\mu_0} \end{bmatrix}. \tag{26}$$

Differentiation of Eq. (5) with respect to $r$ and extension of the definition of the unknown functions, $p(r)$ and $q(r)$, into the range $-a \le r \le 0$ yields.

$$m_1(r) = \frac{1}{2}\int_{-a}^{a} \{p(t)|t|I_{11}(r,t) + q(t)|t|I_{12}(r,t)\}dt + \frac{\alpha_1}{\pi}\int_{-a}^{a} \frac{p(t)}{t-r}dt + \frac{\alpha_1}{\pi}\int_{-a}^{a} p(t)H_1(r,t)dt - \alpha_2 q(r) \tag{27a}$$

$$m_2(r) = \frac{1}{2}\int_{-a}^{a} \{q(t)|t|I_{22}(r,t) + p(t)|t|I_{21}(r,t)\}dt + \frac{\alpha_1}{\pi}\int_{-a}^{a} \frac{q(t)}{t-r}dt + \frac{\alpha_1}{\pi}\int_{-a}^{a} q(t)H_2(r,t)dt + \alpha_2 p(r) \tag{27b}$$

where

$$m_1(r) = \frac{\partial u_{z0}(r,h_0)}{\partial r}, m_2(r) = \frac{1}{r}\frac{\partial r u_{r0}(r,h_0)}{\partial r}, H_i(r,t) = (h_i(r,t)-1)/(t-r),$$

$$I_{ij}(r,t) = (-1)^i \int_0^\infty (sM_{ij}(s,h_0) - \alpha_{ij})sJ_{2-i}(sr)J_{j-1}(st)ds, (i = 1,2, j = 1,2),$$

$$h_1(r,t) = \begin{cases} |t/r|E(t/r), & (|t| < |r|) \\ (t^2/r^2)E(r/t) - \{(t^2 - r^2)/r^2\}K(r/t), & (|t| > |r|) \end{cases},$$

$$h_2(r,t) = \begin{cases} \{(t^2 - r^2)/|tr|\}K(t/r) + (r/t)E(t/r), & (|t| < |r|) \\ E(r/t), & (|t| > |r|) \end{cases},$$

with $K(.)$ and $E(.)$ being, respectively, the complete elliptic integrals of the first and second kinds.

The system of the singular integrals, Eqs. (27a) and (27b), must be solved subjected to the following condition:

$$P = \pi \int_{-a}^{a} p(t)tdt \tag{28}$$

### 4.1 Frictionless contact problem of FGM coating

In this section, the axisymmetric frictionless contact problem between FGM coatings and a rigid spherical punch is studied. As shown in **Figure 4**, an applied force $P$ is acted on the rigid spherical punch along the $z$-direction to form an indent depth $\delta_0$ and a circular contact region with a radius $a$. The displacement boundary condition in the contact region is expressed as

$$w(r,h_0) = \delta_0 - r^2/2R \ (0 \leq r \leq a) \tag{29}$$

Because the frictionless contact is considered, the shear traction $q(r)$ is zero, and the controlling equation is

$$m_1(r) = \frac{1}{2}\int_{-a}^{a} p(t)|t|I_{11}(r,t)dt + \frac{\alpha_1}{\pi}\int_{-a}^{a} \frac{p(t)}{t-r}dt + \frac{\alpha_1}{\pi}\int_{-a}^{a} p(t)H_1(r,t)dt \tag{30}$$

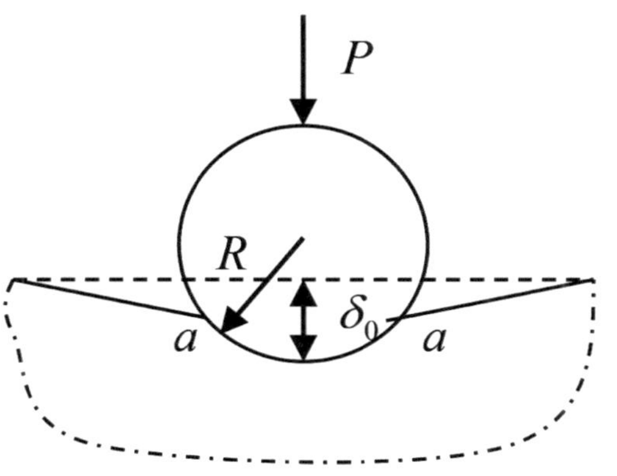


**Figure 4.**
*FGM coating indented by a spherical indenter.*

The Gauss-Chebyshev integration formula [24] is applied to solve Eqs. (28) and (30) with the consideration of Eq. (29).

## 4.2 Partial slip contact problem with finite friction for FGM coating

Consider the axisymmetric partial slip contact problem as shown in **Figure 5**. The normal surface displacement, $u_{z0}$, along the coating interface, $z = h_0$, is given by

$$u_{z0}(r) = \delta_0 - r^2/2R \tag{31}$$

The inner stick region, $r \leq b$, and outer slip annulus, $b \leq r \leq a$, are shown in **Figure 5**. According to Spence's work [18], the radial displacement along the coating interface in the stick region may be expressed as

$$u_{r0}(r) = Cr^2, (r \leq b) \tag{32}$$

where $C$ denotes the slop of the relative radial displacement gradient and is an unknown constant. The Coulomb friction law is applied to describe the slip behavior in the slip region. Then, the radial shear traction in the contact region is represented as

$$q(r) = q*(r) - fp(b)\frac{r}{b}, (r \leq b), \tag{33a}$$

$$q(r) = -fp(r), (b \leq r \leq a). \tag{33b}$$

where $f$ denotes the friction coefficient.

Finally, the partial slip contact problem with consideration of the boundary conditions (31), (32), and (33) can be expressed according to the singular integral equations:

$$-\alpha_2 q(r) + \frac{\alpha_1}{\pi}\int_{-a}^{a}\frac{p(t)}{t-r}dt + \frac{\alpha_1}{\pi}\int_{-a}^{a}p(t)H_1(r,t)dt + \frac{1}{2}\int_{-a}^{a}\{p(t)|t|I_{11}(r,t) + q(t)|t|I_{12}(r,t)\}dt$$
$$= -r/R, \tag{34a}$$

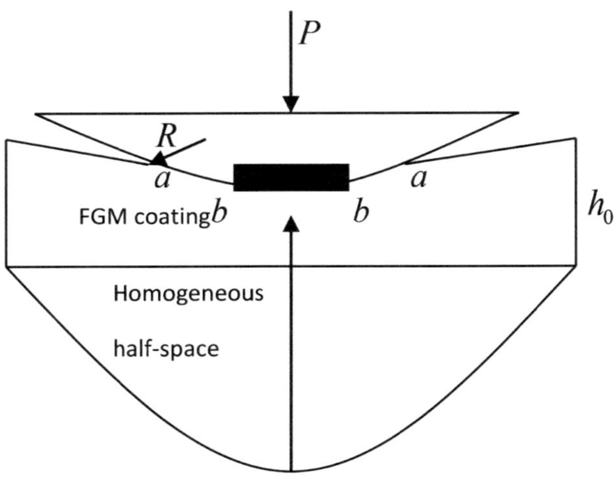


**Figure 5.**
*A functionally graded coated half-space indented by a spherical indenter.*

$$\alpha_2 p(r) + \frac{\alpha_1}{\pi}\int_{-a}^{a}\frac{q(t)}{t-r}dt + \frac{\alpha_1}{\pi}\int_{-a}^{a} q(t)H_2(r,t)dt + \frac{1}{2}\int_{-a}^{a}\{q(t)|t|I_{22}(r,t) + p(t)|t|I_{21}(r,t)\}dt$$
$$= 3C|r| \tag{34b}$$

The Goodman approximate method (uncoupled solution) [21] and the iteration method (coupled solution) [13] are used to solve coupled singular integral equation (34).

## 5. Indentation response of FGM coating under a spherical indenter

The indentation response of FGM coating under frictionless and frictional condition will be presented in this section.

Firstly, the effects of the stiffness ratio $\mu_0/\mu*$ on the distributions of the contact pressure and the relation between indentation and applied force are investigated for the frictionless contact problem. The exponential function model is applied to obtain the results shown in **Figures 6** and 7 [7]. The distribution of the dimensionless contact pressure $p(r)$ (a) and radial stress $\sigma_{rr}(r)$ (b) on the surface of FGM coating indented by a rigid spherical indenter for various stiffness ratio $\mu_0/\mu*$ when $R/h_0 = 10$ and $a/h_0 = 0.2$ is shown in **Figure 6**. With the increase of $\mu_0/\mu*$, the contact pressure $p(r)$ decreases. It can be observed that the tensile spike in the distribution of $\sigma_{rr}(r)$ as $r \to a$ has clearly some implications regarding the initiation and subcritical growth of surface cracks. **Figure 7** presented the relation of $P$ vs. $a$ and $P$ vs. $\delta_0$. With the decrease of $\mu_0/\mu*$, the larger applied normal load is needed to create the same contact region (a) and the same maximum indentation depth $\delta_0$ (b). The results give an indentation testing method to measure the stiffness of the coating surface and the gradient of the coating.

Secondly, the linear multi-layered model is used to model the shear modulus of the coating varying in the following power law form:

$$\mu(z) = \mu* + (\mu_0 - \mu*)(z/h_0)^n, \tag{35}$$

where $n$ is a gradient index characterizing the gradual variation of the shear modulus. In the following calculation, the LML model divided the FGM coating into

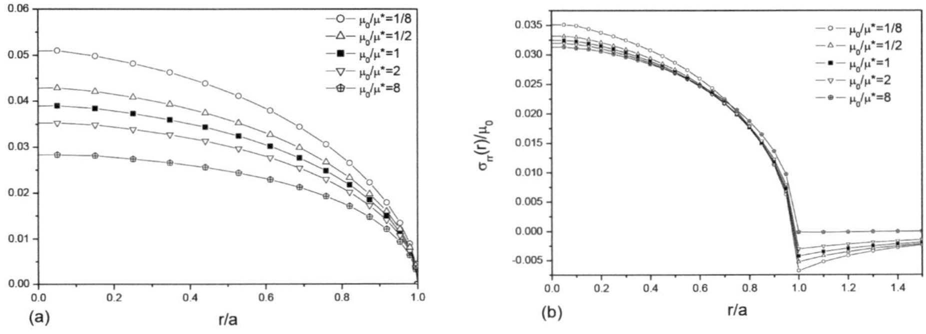


**Figure 6.**
*Distribution of the dimensionless contact pressure $p(r)$ (a) and radial stress $\sigma_{rr}(r)$ (b) on the surface of the graded coating loaded by a rigid spherical indenter for some selected values of the stiffness ratio $\mu_0/\mu*$ with $R/h_0 = 10$ and $a/h_0 = 0.2$.*

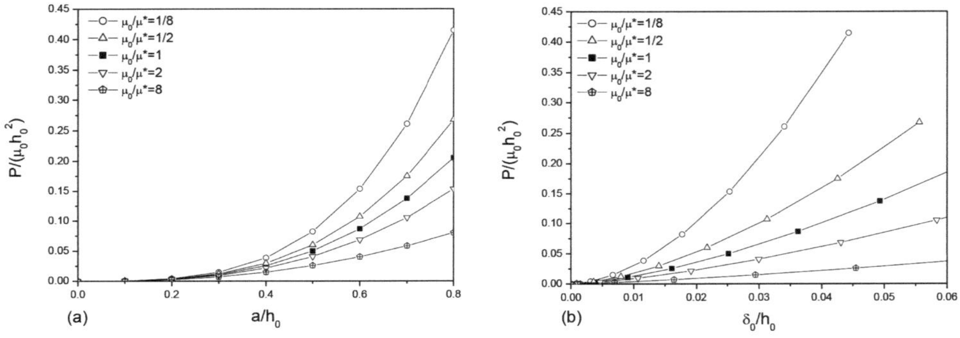


**Figure 7.**
*Relations of P vs. a (a) and P vs. $\delta_o$ (b) for some selected values of the stiffness ratio $\mu_o/\mu*$ with $R/h_o = 10$.*

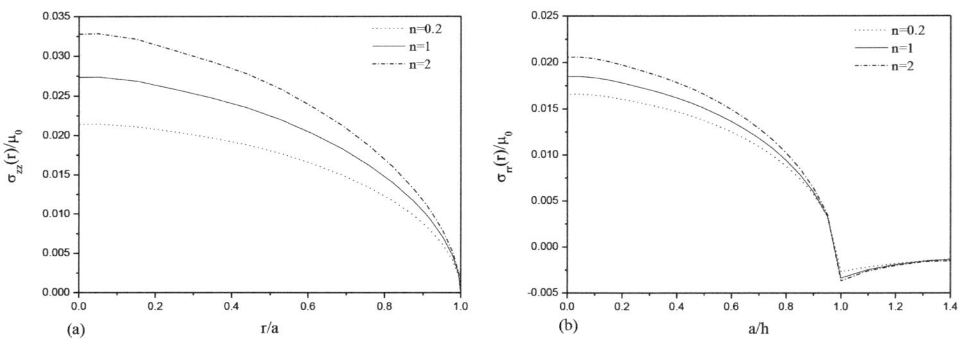


**Figure 8.**
*Distribution of the dimensionless contact pressure $\sigma_{zz}(r)$ (a) and radial stress $\sigma_{rr}(r)$ (b) for some selected values n with $a/h = 0.1$ and $R/h_o = 10$.*

six sub-layers. The axisymmetric indentation response for the frictionless contact under the spherical indenter is considered.

**Figure 8** shows the distributions of the contact pressure for some selected values of n with $\mu_0/\mu*$ = 1/8 and $a/h_0 = 0.1$ [12]. With the increase of $n$, the contact pressure obviously increases. This behavior shows that the contact traction can be improved by adjusting the gradient of the coating when the stiffness of the coating surface keeps unchanged. When the FGM coating is indented by a conical indenter, the relations of $P$ vs. $a$ (a) and $P$ vs. $\delta_0$ (b) for some selected values of $n$ with $\mu_0/\mu*$ = 1/8 are shown in **Figure 9** [12]. To create the same contact region and the

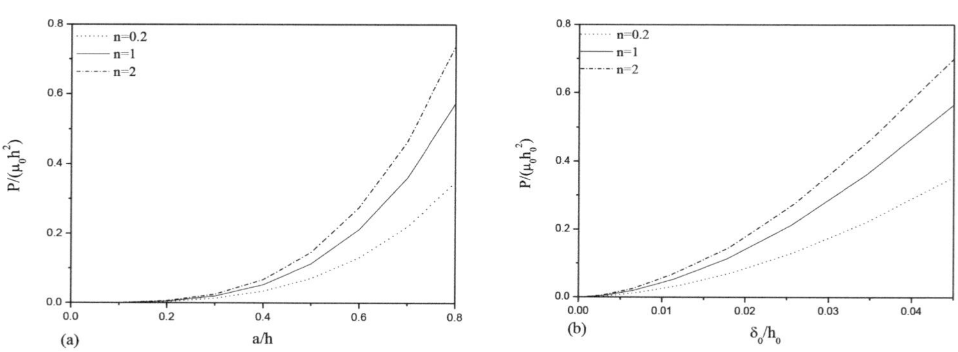


**Figure 9.**
*Relations of P vs. a (a) and P vs. $\delta_o$ (b) for selected values of n with $R/h = 10$.*

same maximum indentation depth $\delta_0$ (b), the larger applied normal load is needed for larger values of $n$.

Thirdly, the effect of the variation of Poisson's ratio on the frictionless contact problem is considered by using piece wise exponential multi-layered model. The shear modulus of FGM coating varies as power law form according to Eq. (35). Poisson's ratio of FGM coating is assumed to vary as the linear function along the thickness as follows

$$v_j = v_1 + \frac{j-1}{N-1}(v* - v_1), j = 1, 2, \ldots\ldots N \tag{36}$$

where $v_1$ and $v*$ are Poisson's ratio for the first layer and homogeneous half-space. $v_j$ denotes Poisson's ratio in layer $j$. The contact pressure (a) and the relations of $P$ vs. $a$ (b) for the different variation forms of Poisson's ratio when $v_1$ = 1/3 and $\mu_0/\mu*$ = 1/5 are given in **Figure 10** [14]. It is assumed that Poisson's ratio for the FGM coating-substrate structure varies from 1/3 to 0.1 and varies from 1/3 to 0.5 according to Eq. (36). The results show that the variation of Poisson's ratio along the thickness has no significant impact on the contact pressure and the relation of force contact region in axisymmetric contact problem when the Poisson's ratio at the upper surface of coating is fixed. **Figure 11** presented the effect of value of Poisson's ratio on the contact pressure (a) and the relation of $P$ vs. $a$ (b) when Poisson's ratio in the coating-substrate structure is a constant ($v_j = v$) as shown in [14]. We can observe that the value of Poisson's ratio has a significant effect on the contact

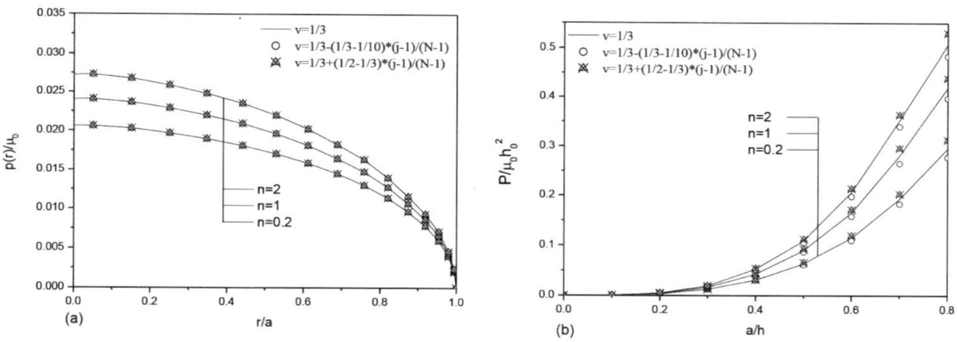


**Figure 10.**
*The contact pressure (a) and the relations of P vs. a (b) for the different variation form of the Poisson's ratio with* $v_1$ = 1/3 *and* $\mu_0/\mu*$ = 1/5.

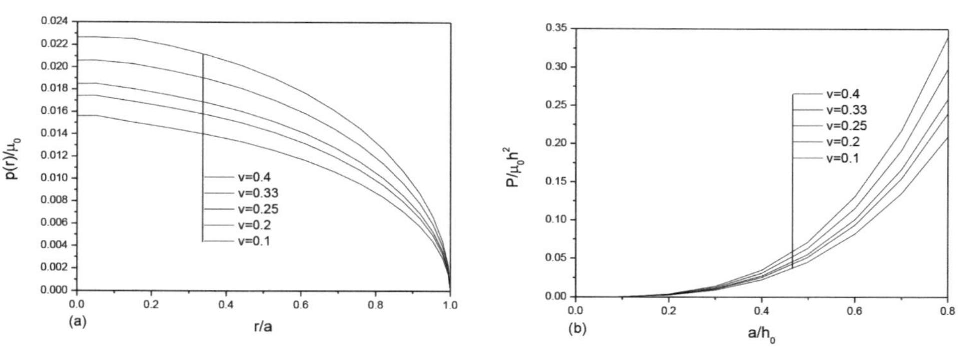


**Figure 11.**
*The effect of the value of Poisson's ratio on the contact pressure (a) and on the relation of P vs. a (b) with* $\mu_0/\mu*$ = 1/5 *while n* = 0.2.

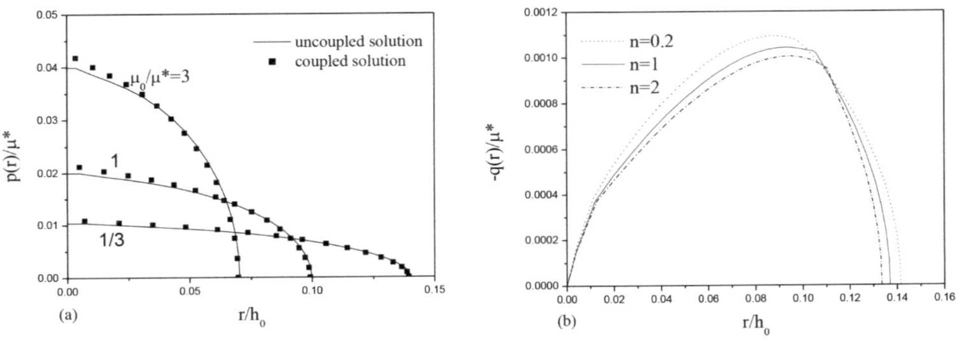


**Figure 12.**
*Contact traction distributions for selected values of the shear modulus ratio* $\mu_0/\mu*$ *with* $P/\mu * h^o{}_2 = 4 \times 10^{-4}$ *and* $f = 0.16$*: (a)* $p(r)$ *and (b)* $q(r)$.

pressure. While the values of $v$ obviously increase, the contact pressure is observed. The results also show that the larger applied normal load is needed to create the same contact region and the same maximum indentation depth $\delta_0$ for larger values of $v$.

Finally, the axisymmetric contact problem of a functionally graded coated half-space is indented by a rigid spherical punch in the case of the partial slip. The linear multi-layered model is used to solve the problem.

The normal contact traction and radial tangential traction for some selected values of the shear modulus ratio $\mu_0/\mu*$ with $P/\mu * h^0{}_2 = 4 \times 10^{-4}$ and $f = 0.16$ are shown in **Figure 12** [14]. The solid lines correspond to the uncoupled solution, and the scatter symbols correspond to the coupled solution. We can observe that consideration of the coupling between the normal and tangential tractions may result in the increase of the peak contact tractions but slight decrease of the contact tractions near the edges of the contact region for a given shear modulus ratio $\mu_0/\mu*$. With the increase of $\mu_0/\mu*$, the peak normal and tangential contact tractions increase. **Figure 12b** also shows that the stick region and the contact radius decrease with the increase of $\mu_0/\mu*$. This behavior provides a way for us to change the distribution of the contact pressure by adjusting the stiffness of the coating surface. **Figure 13** presents the effects of $n$ on the contact traction distributions with $P/\mu * h^0{}_2 = 4 \times 10^{-4}$ and $f = 0.16$ [14]. With the increase of $n$, the peak normal traction (**Figure 13a**) increases, and the peak tangential traction (**Figure 13b**) decreases. This behavior provides a way for us to change the distribution of the

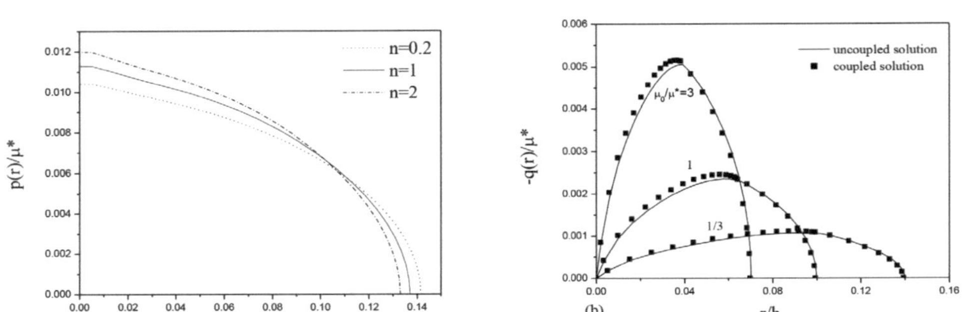


**Figure 13.**
*Contact traction distributions for selected values of n with* $P/\mu * h^o{}_2 = 4 \times 10^{-4}$ *and* $f = 0.16$*: (a)* $p(r)$ *and (b)* $q(r)$.

contact traction by adjusting the gradient of the coating while remaining the shear modulus of the coating surface unchanged.

## 6. Conclusions

In this chapter, we introduced the axisymmetric indentation response for FGM coating under frictionless and partial slip condition by using the three types of computational models. The exponential function model can solve the axisymmetric contact problem for FGM coating whose elastic modulus continuously varies, but it cannot simulate FGM with arbitrarily varying properties. The linear multi-layered model allows arbitrarily the variation of the material properties of FGM, but it requires Poisson's ratio which is 1/3. The Piece wise exponential multi-layered model can simulate functionally graded coating with arbitrarily varying material modulus with no limit to Poisson's ratio, but numbers of sub-layers are larger. In practice, the computational model is chosen according to properties of the problem. Hankel integral transformation technology and transfer matrix method are used to solve the axisymmetric contact problem of FGM coating based on the cylindrical coordinate system. The results show that the contact behavior can be improved by adjusting the gradient of FGM coating. The present investigation will be expected to provide a guidance for design considerations and applications of FGM coating.

## Acknowledgements

The support from the National Natural Science Foundation of China under Grant Nos. 11662011 and 11811530067 are gratefully acknowledged.

## Author details

Tie-Jun Liu
Department of Mechanics, Inner Mongolia University of Technology, Hohhot, China

*Address all correspondence to: liutiejun6204@163.com

## References

[1] Yamanoushi M, Koizumi M, Hirai T, Shiota I, editors. Proceedings of the First International Symposium on Functionally Gradient Materials. Sendai, Japan; 1980

[2] Suresh S. Graded materials for resistance to contact deformation and damage. Science. 2001;**292**:2447-2451

[3] Giannakopoulos AE, Suresh S. Indentation of solids with gradients in elastic properties: Part I. Point force solution. International Journal of Solids and Structures. 1997;**34**:2357-2392

[4] Giannakopoulos AE, Suresh S. Indentation of solids with gradients in elastic properties: Part II. Axisymmetric indenters. International Journal of Solids and Structures. 1997;**34**:2392-2428

[5] Guler MA, Erdagon F. Contact mechanics of graded coatings. International Journal of Solids and Structures. 2004;**41**:3865-3889

[6] Guler MA, Erdagon F. Contact mechanics of two deformable elastic solids with graded coatings. Mechanics of Materials. Cambridge, UK: 2006;**38**: 633-647

[7] Liu TJ, Wang YS. Axisymmetric frictionless contact problem of a functionally graded coating with exponentially varying modulus. Acta Mechanica. 2008;**199**:151-165

[8] Liu TJ, Xing YM. The axisymmetric contact problem of a coating/substrate system with a graded interfacial layer under a rigid spherical punch. Mathematics and Mechanics of Solids. 2016;**21**:383-399

[9] Jeon SP, Tanigawa Y, Hata T. Axisymmetric problem of a nonhomogeneous elastic layer. Archive of Applied Mechanics. 1998;**68**:20-29

[10] Ke LL, Wang YS. Two-dimensional contact mechanics of functionally graded materials with arbitrary spatial variations of material properties. International Journal of Solids and Structures. 2006;**4**:5779-5798

[11] Ke LL, Wang YS. Two-dimensional sliding frictional contact of functionally graded materials. European Journal of Mechanics - A/Solids. 2007;**26**:171-188

[12] Liu TJ, Wang YS, Ch Z. Axisymmetric contact problem of functionally graded materials. Archive of Applied Mechanics. 2008;**78**:267-282

[13] Liu TJ, Wang YS, Xing YM. The axisymmetric partial slip contact problem of a graded coating. Meccanica. 2012;**47**:1673-1693

[14] Liu TJ, Xing YM. Analysis of graded coatings for resistance to contact deformation and damage based on a new multi-layer model. International Journal of Mechanical Sciences. 2014;**81**: 158-164

[15] Liu TJ, Li PX. Two-dimensional adhesion mechanics of a graded coated substrate under a rigid cylindrical punch based on a PWEML model. Applied Mathematical Modelling. 2019; **69**:1-14

[16] Hertz H. On the Contact of Elastic Solids. Miscellaneous Papers by H. Hertz. London, UK: Macmillan; 1882

[17] Johnson KL. Contact Mechanics. UK: Cambridge University Press; 1985

[18] Spence DA. The hertz contact problem with finite friction. Journal of Elasticity. 1975;**5**:297-319

[19] Ke LL, Wang YS. Fretting contact of functionally graded coated half-space with finite friction—Part I—Normal

loading. Journal of Strain Analysis for Engineering Design. 2007;**42**:293-304

[20] Spence DA. Self-similar solutions to adhesive contact problems with incremental loading. Proceedings of the Royal Society of London. Series A: Mathematical and Physical Sciences. 1968;**305**:55-80

[21] Goodman LE. Contact stress analysis of normally loaded rough spheres. Journal of Applied Mechanics. 1962;**29**:515-522

[22] Mossakovski VI. Compression of elastic bodies under condition of adhesion (axisymmetric case). Szhatie uprugikh tel v usloviiakh stsepleniia (Osesimmetbichnyi sluchai): PMM. Journal of Applied Mathematics and Mechanics. 1963;**27**:418-427

[23] Norwell D, Hills DA, Sackfield A. Contact of dissimilar elastic cylinders under normal and tangential loading. Journal of the Mechanics and Physics of Solids. 1988;**36**:59-75

[24] Civelek MB. The axisymmetric contact problem for an elastic layer on a frictionless half-space [thesis]. Mechanical Engineering Department Lehigh University; 1972